Elementary Geometry of Differentiable Curves:
an Undergraduate Introduction

Elementary Geometry of Differentiable Curves: an Undergraduate Introduction

C. G. Gibson

PUBLISHED BY THE PRESS SYNDICATE OF THE UNIVERSITY OF CAMBRIDGE
The Pitt Building, Trumpington Street, Cambridge, United Kingdom

CAMBRIDGE UNIVERSITY PRESS
The Edinburgh Building, Cambridge CB2 2RU, UK
40 West 20th Street, New York, NY 10011–4211, USA
10 Stamford Road, Oakleigh, VIC 3166, Australia
Ruiz de Alarcón 13, 28014 Madrid, Spain
Dock House, The Waterfront, Cape Town 8001, South Africa

http://www.cambridge.org

© Cambridge University Press 2001

This book is in copyright. Subject to statutory exception
and to the provisions of relevant collective licensing agreements,
no reproduction of any part may take place without
the written permission of Cambridge University Press.

First published 2001

Printed in the United Kingdom at the University Press, Cambridge

Typeface Times 10/13pt.   *System* LaTeX 2_ε [TYPESET BY THE AUTHOR]

A catalogue record for this book is available from the British Library

Library of Congress Cataloguing in Publication data

ISBN 0 521 80453 1 hardback
ISBN 0 521 01107 8 paperback

To Dorle

Contents

List of Illustrations		*page* x
List of Tables		xiii
Preface		xiv
1	**The Euclidean Plane**	1
1.1	The Vector Structure	1
1.2	The Scalar Product	2
1.3	Length, Distance and Angle	2
1.4	The Complex Structure	6
1.5	Lines	7
1.6	Projection to Lines	11
2	**Parametrized Curves**	13
2.1	The General Concept	13
2.2	Self Crossings	20
2.3	Tangent and Normal Vectors	23
2.4	Tangent and Normal Lines	27
3	**Classes of Special Curves**	32
3.1	The Standard Conics	32
3.2	General Algebraic Curves	38
3.3	Trochoids	42
4	**Arc Length**	50
4.1	Arc Length	50
4.2	Parametric Equivalence	53
4.3	Unit Speed Curves	57
4.4	Involutes	58
5	**Curvature**	62
5.1	The Moving Frame	62
5.2	The Concept of Curvature	64
5.3	Calculating the Curvature	65

5.4	Inflexions	69
5.5	Limiting Behaviour	73
6	**Existence and Uniqueness**	75
6.1	Isometries	76
6.2	Fixed Points and Formulas	79
6.3	Congruent Curves	82
6.4	The Uniqueness Theorem	85
7	**Contact with Lines**	89
7.1	The Factor Theorem	89
7.2	Multiplicity of a Zero	90
7.3	Contact with Lines	92
7.4	Inflexions and Undulations	94
7.5	Cusps	99
8	**Contact with Circles**	105
8.1	Contact Functions	105
8.2	Evolutes	109
8.3	Parallels	119
9	**Vertices**	124
9.1	The Concept of a Vertex	124
9.2	Appearance of Vertices on the Evolute	131
9.3	The Four Vertex Theorem	133
10	**Envelopes**	137
10.1	Envelopes	137
10.2	The Envelope Theorem	140
10.3	Natural Envelopes in Geometry	146
11	**Orthotomics**	151
11.1	Reflexions	151
11.2	Orthotomics	153
11.3	Orthotomics of Non-Regular Curves	160
11.4	Irregular Points on Orthotomics	162
11.5	Antiorthotomics	162
12	**Caustics by Reflexion**	165
12.1	Caustics of a Curve	166
12.2	Caustics as Evolutes	169
12.3	Sources at Infinity	173
12.4	Orthotomics as Envelopes	177
13	**Planar Kinematics**	180
13.1	Historical Genesis	180
13.2	Planar Motions	184
13.3	General Roulettes	186

14	**Centrodes**	190
14.1	Generic Parameters	190
14.2	Generic Parameters for Roulettes	192
14.3	Fixed and Moving Centrodes	194
15	**Geometry of Trajectories**	199
15.1	Equivalence of Motions	200
15.2	Cusps on Trajectories	203
15.3	Inflexions on Trajectories	205
15.4	Vertices on Trajectories	209
Index		211

List of Illustrations

1.1	Components of a vector	3
1.2	The two sides of a line	9
1.3	The orthogonal bisector	10
1.4	Three ways in which lines can intersect	11
1.5	Orthogonal projection onto a line	12
2.1	A parametrized curve	14
2.2	The right strophoid	15
2.3	The catenary	16
2.4	Parametrizing the circle	18
2.5	Some rose curves	19
2.6	Tschirnhausen's cubic	20
2.7	The eight-curve	21
2.8	Cayley's sextic	22
2.9	Tangent and normal vectors	23
2.10	The semicubical parabola	24
2.11	Construction of the piriform	25
2.12	The piriform	26
2.13	Curve with control points b_0, b_1, b_2, b_3	27
2.14	The cross curve	28
2.15	The tractrix	29
3.1	Parabola as a standard conic	33
3.2	Ellipse as a standard conic	35
3.3	Hyperbola as a standard conic	36
3.4	Agnesi's versiera	39
3.5	The cissoid of Diocles	40
3.6	The idea of a roulette	42
3.7	Construction of trochoids	43
3.8	Various epicycloids and hypocycloids	45
3.9	Three different types of limacon	46
3.10	Construction of cycloids	47

3.11	Forms of the cycloid	48
4.1	The astroid	51
4.2	The idea of parametric equivalence	54
4.3	Some Lissajous figures	56
4.4	The involute construction	59
4.5	An involute of a circle	60
5.1	The idea of a moving frame	63
5.2	A typical inflexion	69
5.3	A biflexional limacon	72
5.4	Forms of the curves near the origin	74
6.1	Superimposing two curves	76
6.2	Congruent curves	82
6.3	An equiangular spiral	83
7.1	Curve crossing a non-tangent line L	94
7.2	Picturing the sign of the curvature	96
7.3	Inflexions having odd and even contact	97
7.4	The Serpentine	98
7.5	A cuspidal tangent line	100
7.6	Example of a higher cusp	102
8.1	Circles tangent at the vertex of a parabola	107
8.2	A parabola and its evolute	110
8.3	An ellipse and its evolute	111
8.4	The evolute of the eight-curve	112
8.5	Three cusps and their evolutes	113
8.6	Evolutes of some epicycloids and hypocycloids	115
8.7	A cycloid and its evolute	116
8.8	The tractrix and its evolute	117
8.9	Cayley's sextic and its evolute	118
8.10	Parallels of a parabola	120
8.11	Parallels of an ellipse	121
8.12	Two parallels of an astroid	122
9.1	Vertices on the graph of $f(x) = x^2(2x + 3)$	128
9.2	A cardioid and its evolute	132
9.3	Convex and non-convex ovals	134
9.4	Proof of Lemma 9.3	135
10.1	A family of curves	138
10.2	A family of circles	138
10.3	Two envelopes of the family of circles	139
10.4	Astroid as an envelope of ellipses	142
10.5	Nephroid as an envelope of circles	143
10.6	Cardioid as an envelope of lines	145
10.7	Parabola parallel as a circle envelope	147
10.8	Envelope of the normals for a parabola	148

10.9	Envelope of the normals for an ellipse	149
11.1	Construction of the orthotomic	153
11.2	Three orthotomics of a circle	155
11.3	Orthotomics of a parabola	156
11.4	Orthotomic of an ellipse with source a focus	157
11.5	Bernoulli's lemniscate	158
11.6	Orthotomics of some hypocycloids	161
12.1	Coffee cup caustic	165
12.2	Caustic by reflexion	166
12.3	Constructing the caustics of a circle	167
12.4	Caustics by reflexion for a circle	168
12.5	The reflective property for an ellipse	172
12.6	Orthotomic with respect to a line	174
12.7	Orthotomic of a circle with source at infinity	175
12.8	Orthotomic of a parabola with source at infinity	176
12.9	The reflective property for a parabola	177
12.10	Orthotomic as an envelope of circles	178
13.1	The Watt four bar	181
13.2	Four bar model	182
13.3	A four bar trajectory	183
13.4	Collapsible four bar linkages	183
13.5	The parallelogram four bar	184
13.6	The double slider	185
13.7	The idea of a roulette	186
14.1	Centrodes of the double slider	195
14.2	Crossed parallelogram four bars	196
14.3	Centrode for a crossed parallelogram	197

List of Tables

3.1	Special epicycloids and hypocycloids	44
7.1	Inflexions on limacons	98
8.1	Evolutes of some epicycloids and hypocycloids	114
11.1	Orthotomics of epicycloids and hypocycloids	161

Preface

For around three decades one of the distinguishing features of my department† has been a second year course on the geometry of curves, in which (following an earlier set of precepts) plane curves are studied simultaneously from the algebraic and differentiable viewpoints. It has a proven history of success, providing students with a wonderful introduction to visually attractive geometry. My experience in teaching this course convinced me that the time was ripe to raise the profile of undergraduate geometry. The algebraic viewpoint developed into my text, 'Elementary Geometry of Algebraic Curves' (Cambridge University Press, 1998). The present text is intended as a companion volume, representing the differentiable viewpoint.

0.1 Differential Geometry

Differential geometry is a fascinating area of mathematics, of substantial and increasing importance in the physical sciences. Despite that, the subject has a low billing in most undergraduate curricula, either not appearing or relegated to a final year optional course. That is a shame since much can be achieved with minimal mathematical preparation in the second year by restricting attention to plane curves: those who then wish to develop their interest can proceed to final year courses studying more general objects. Plane curves live in an environment familiar from school mathematics (the Euclidean plane), and have features readily visible on a computer screen: moreover, their study uses foundational mathematics (calculus, linear algebra and complex numbers) in a useful way, with only a handful of results using basic facts in real analysis. Most

† The Department of Mathematical Sciences in the University of Liverpool

Preface

computations use no more than elementary symbolic manipulation and differentiation, thus reinforcing the basic skills acquired in school and the first year of undergraduate study. The restriction to the plane has the further benefit that its natural complex structure can be profitably exploited to simplify both theory and practice.

0.2 Special Curves

One of my objectives was to ensure that the reader becomes increasingly familiar with a small zoo of special curves, mostly drawn from physical applications, to provide a framework on which the ideas of the subject can be hung. Many of the curves of historical significance can be studied profitably from either the algebraic or the differentiable viewpoint, so provide a useful starting point. Indeed it is healthy to bear in mind that the study of such special curves provided the genesis for the existing body of mathematics. Moreover some of these curves exhibit subtle features, requiring careful analysis. I resisted the temptation to expand upon the historical aspects: it would require a separate volume, and an author with much wider knowledge. However, most of these named curves are either of historical, or of mathematical significance – sometimes both. For instance the catenary was discovered by Galileo (who confused it with the parabola) and later studied by Jacques Bernoulli the Elder (who discovered its true form). But it is also of mathematical significance, as a plane section of the minimal surface spanning two circular discs, the only minimal surface of revolution.

0.3 Curvature, Contact and Caustics

Curvature is of course one of the central concepts. Here is one of the big ideas of mathematics, compelling in its simplicity, yet surprisingly subtle. In time honoured fashion it is shown that a curve is completely determined (up to congruence) by its speed and curvature functions. This seems to me to be an excellent illustration of Klein's 'Erlanger Programm', one of the few accessible to undergraduates prior to their final year of study. I wanted to ensure that students also had the opportunity of viewing elementary differential geometry from the (less familiar) singular viewpoint. Thus two chapters are devoted to studying the contact of curves with lines and circles, leading to an understanding of exceptional points on curves such as cusps, inflexions and vertices. The singular theme is continued via a discussion of envelopes. As a serious application, two chapters are devoted to caustics by reflexion, an attractive area of potential application to several areas of the physical sciences, which deserves to be better known. Caustics are already of considerable importance in geometric optics. However, their significance in acoustics is not so well established, and there is little doubt that they play a key role in understanding the mysteries of radio propagation.

0.4 Planar Kinematics

The later chapters develop a subject close to my heart, namely kinematics. Like my engineering colleagues, I regret the fact that this area of study has largely disappeared from mathematics degrees. It deserves putting into a historical context, for the subject is as old as mathematics itself. It relates intimately to two of the great social revolutions – the Power Revolution, when man was gradually released from the drudgery of providing a source of power (as ways became available of converting natural sources of power into mechanical work) and the subsequent Industrial Revolution. Planar motions with a single degree of freedom represent the core material of classical kinematics: it stands on its own as interesting geometry, having intimate and fruitful relations with other areas of mathematics. The simplest examples arise from the roulettes of curve theory, indeed it is one of the key results of the subject that general planar motions arise in this way. (It is for that reason that trochoids, curves traced by a point carried by one circle rolling on another, are introduced at an early stage.) Here again the singular viewpoint is of historical importance, producing the circle of inflexions and the cubic of stationary curvature, sadly far better known to engineers than mathematicians. And, looking to the future, one sees this classical core as the starting point of wider theoretical investigations into robotics, a subject which plays an ever increasing role in our everday lives.

0.5 Concerning the Structure

In keeping with my objective of writing for undergraduates, with a year of foundational mathematics behind them, this volume is unashamedly example based. It is my firm personal belief that there is much educational merit in really getting to grips with a range of explicit examples. The subject is indeed rich in attractive examples, many of which arose in the physical sciences, and are of historical significance. I believe that students gain in confidence from this approach, and enjoy the security of increasing familiarity with key examples. Those who wish to can always pursue abstract theory for its own sake by proceeding to higher degrees where their needs will be met. The format parallels that of the companion volume 'Elementary Geometry of Algebraic Curves'. I kept the individual chapters fairly short, on the theory that each chapter revolves around one new idea: likewise the sections are brief, and punctuated by a series of 'examples' illustrating the concepts. I included a substantial and coherent collection of exercises (many culled from the older literature) designed to illustrate (and even amplify) the small amount of theory. Each chapter contains sets of exercises, each appearing immediately after the relevant section.

0.6 Curve Tracing

A few words are in order on the subject of curve tracing. At the crudest level, there is a lot to be said for a thumbnail sketch of a curve to grasp the broadest qualitative features. Many interesting curves arise from simple geometric constructions, and can be traced on paper using no more than a school geometry set. The serious tracer will enjoy Lockwood's 'A Book of Curves' where such constructions are described in considerable detail. It is only one step further in this direction to acquire a spirograph, and enjoy the sheer beauty of the complex trochoids which it will trace. And if that does not satisfy you, seek out such gems as Alabone's gorgeous Edwardian collection† of colour illustrations produced by an intriguing mechanical device, the Epicycloidal Geometric Chuck! I mourn the decline in such scientific hobbies: however, those who feel that life is too short for such indulgences will find that computers provide superb renderings of curves in a fraction of a second. It is worth saying that curve tracing programs allow both student and teacher to experiment with curves interactively on a computer screen, thereby enhancing understanding of the underlying geometry. Most commercial mathematical software packages contain such programs: indeed many of the illustrations in this book were produced in MAPLE. The more adventurous will not find it difficult to write simple programs themselves illustrating individual curves, and even families of curves.

---

† 'Poly-cyclo-epicycloidal and other Geometric Curves'

1
The Euclidean Plane

We need to lay firm foundations for later work. It is rather like gardening. We would like to have beautiful flowers appear immediately, but as gardeners well know, success is the result of careful preparation of the soil, and a generous measure of patience. So it is in mathematics. The soil in which our curves grow is the familiar plane of school geometry. Good preparation now will make our lives much easier at a later stage. Virtually everything in this book depends crucially on the 'Euclidean structure' of the plane. That provides the content of this introductory chapter, representing purely foundational material. For many readers this material will already be a part of their knowledge, in which case it might be best just to scan the contents and proceed to Chapter 2.

1.1 The Vector Structure

Throughout this text $\mathbb{R}$ will denote the set of real numbers. For linguistic variety we will sometimes refer to real numbers as *scalars*. We will work in the familiar real plane $\mathbb{R}^2$ of elementary geometry, whose elements $z = (x, y)$ are called *points* (or *vectors*). Recall that we can add vectors, and multiply them by scalars λ, according to the familiar rules

$$(x_1, y_1) + (x_2, y_2) = (x_1 + x_2, y_1 + y_2)$$
$$\lambda(x, y) = (\lambda x, \lambda y).$$

Two vectors $z_1 = (x_1, y_1)$, $z_2 = (x_2, y_2)$ are *linearly independent* when $x_1 y_2 - x_2 y_1 \neq 0$. Linear algebra tells us that if z_1, z_2 are linearly independent then they form a basis for $\mathbb{R}^2$, meaning that any vector z can be written uniquely in the form $z = \lambda_1 z_1 + \lambda_2 z_2$ for some scalars λ_1, λ_2: in that case we say that z has *coordinates* (λ_1, λ_2) relative to the basis z_1, z_2. The most familiar case is the *standard basis* comprising the vectors

$e_1 = (1, 0)$, $e_2 = (0, 1)$ giving rise to the *standard coordinates* of elementary geometry.

1.2 The Scalar Product

The plane is endowed with its standard *Euclidean structure*. By this we mean that for any two vectors $z_1 = (x_1, y_1)$, $z_2 = (x_2, y_2)$ we have the standard *scalar product* (or *dot product*) defined by the relation

$$z_1 \bullet z_2 = x_1 y_1 + x_2 y_2.$$

The basic properties (Exercise 1.2.1) of the scalar product are listed below.

S1: $z \bullet z \geq 0$ with equality if and only if $z = 0$.
S2: $z \bullet w = w \bullet z$.
S3: $z \bullet (\lambda w) = \lambda(z \bullet w)$.
S4: $z \bullet (w + w') = z \bullet w + z \bullet w'$.

S2 is referred to as the *symmetry* property. Properties S3, S4 together say that $\bullet$ is linear in its second argument: by symmetry, it is also linear in its first argument, and for that reason $\bullet$ is said to be *bilinear*. Two vectors z, w are *orthogonal* when $z \bullet w = 0$.

Example 1.1 Let z, w be vectors with $w \neq 0$. We claim that there is a unique scalar λ with the property that the vectors $z' = z - \lambda w$, w are orthogonal. Indeed, our requirement is that

$$0 = z' \bullet w = (z - \lambda w) \bullet w = z \bullet w - \lambda(w \bullet w),$$

giving the unique solution $\lambda = z \bullet w / w \bullet w$. We call the vector λw the *component of z parallel to w*, and the vector $z' = z - \lambda w$ the *component of z orthogonal to w*.

Exercises

1.2.1 Starting from the definition of the scalar product, establish the properties S1, S2, S3, S4.

1.3 Length, Distance and Angle

Property S1 of the scalar product is expressed by saying that the scalar product is *positive definite*. In view of this property it makes sense to

1.3 Length, Distance and Angle

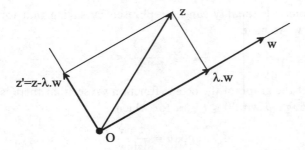

Figure 1.1. Components of a vector

define the *length* of a vector $z = (x, y)$ to be

$$|z| = \sqrt{x^2 + y^2} = \sqrt{z \bullet z}.$$

Throughout this book we will use the following fundamental properties L1, L2, L3 of the length function. The property L1 is an immediate consequence of S1 above: however, L2 and L3 require proof, representing the next step in our development.

L1: $|z| = 0$ if and only if $z = 0$.
L2: $|z \bullet w| \leq |z||w|$. (The Cauchy Inequality.)
L3: $|z + w| \leq |z| + |w|$. (The Triangle Inequality.)

Lemma 1.1 *For any two vectors z, w in the plane we have the relation $|z \bullet w| \leq |z||w|$.* (The Cauchy Inequality.)

Proof When $z = 0$ then the LHS is zero, and the inequality is satisfied. We can therefore assume that $z \neq 0$, so $z \bullet z > 0$. Set $\lambda = z \bullet w / z \bullet z$. Then λw represents the component of z parallel to w, and $z - \lambda w$ is the component orthogonal to w. (Figure 1.1.) Then

$$\begin{aligned} 0 &\leq |w - \lambda z|^2 = (w - \lambda z) \bullet (w - \lambda z) \\ &= w \bullet w - 2\lambda(z \bullet w) + \lambda^2(z \bullet z) \\ &= w \bullet w - \lambda(z \bullet w) \\ &= |w|^2 - \frac{(z \bullet w)^2}{|z|^2}. \end{aligned}$$

The result follows on multiplying through by $|z|^2$ and taking positive square roots. □

The Cauchy Inequality can be rephrased by saying that for non-zero vectors z, w we have

$$-1 \le \frac{z \bullet w}{|z||w|} \le 1.$$

Looking at the graph of the cosine function we see that there is therefore a unique angle θ with $0 \le \theta \le \pi$ for which

$$\cos \theta = \frac{z \bullet w}{|z||w|}.$$

We call θ the *angle* between the vectors z, w. This provides a good intuition for the meaning of the scalar product. The vectors $z = (x, y)$ for which $|z| = 1$ are called *unit* vectors: they are the vectors which lie on the circle $x^2 + y^2 = 1$ of radius 1 centred at the origin. When z, w are unit vectors the scalar product is just the cosine of the angle between them. It is for this reason that we call two vectors z, w 'orthogonal' when $z \bullet w = 0$, since then the cosine is zero, and the angle is $\theta = \pi/2$.

Lemma 1.2 *For any two vectors z, w in the plane we have the relation $|z + w| \le |z| + |w|$. (The Triangle Inequality.)*

Proof The Cauchy Inequality yields the following series of relations, from which the result follows on taking positive square roots:

$$\begin{aligned}
|z+w|^2 &= (z+w) \bullet (z+w) = z \bullet z + 2(z \bullet w) + w \bullet w \\
&\le |z|^2 + 2|z \bullet w| + |w|^2 \le |z|^2 + 2|z||w| + |w|^2 \\
&= (|z| + |w|)^2.
\end{aligned}$$

$\square$

We define the *distance* between two points u, v in the plane to be the scalar $|u - v|$. The following basic properties of the distance function are immediate from L1, L2, L3.

M1: $|u - v| = 0$ if and only if $u = v$.
M2: $|u - v| = |v - u|$.
M3: $|u - v| \le |u - w| + |w - v|$.

Note that distance is *invariant under translation*, in the sense that for any vector w the distance between u, v equals that between their translates $u + w, v + w$.

Example 1.2 Let a, b, c be non-zero vectors with $c = a - b$, and let θ be the angle between a, b. Expanding the expression for $|c|^2$ we obtain the *cosine rule* of school trigonometry

$$|c|^2 = |a|^2 - 2|a||b|\cos\theta + |b|^2.$$

A special case arises when a, b are orthogonal, so the angle θ is a right angle: the cosine rule then reduces to the familiar *Pythagoras Theorem*

$$|c|^2 = |a|^2 + |b|^2.$$

A basis u, v for the Euclidean plane is *orthogonal* when u, v are orthogonal: and it is *orthonormal* when it is orthogonal, and u, v are in addition unit vectors.

Example 1.3 Let T, N be orthogonal unit vectors. (The symbols are chosen deliberately to reflect the geometric situations we will meet throughout this text.) Thus the assumptions are that $T \bullet N = 0$, $T \bullet T = 1$, $N \bullet N = 1$. Observe first that T, N are linearly independent. Suppose indeed that we had a relation $\tau T + \nu N = 0$ for some scalars τ, ν: then, taking scalar products of each side of this relation with T, N, we see that $\tau = 0$, $\nu = 0$. By linear algebra T, N form an orthonormal basis for the plane, so that any vector v can be written uniquely as a linear combination $v = \tau T + \nu N$ for some scalars τ, ν. These scalars are very easily determined. Taking scalar products with T we see that $\tau = v \bullet T$; and likewise, taking scalar products with N we see that $\nu = v \bullet N$. Thus the required linear combination is

$$v = (v \bullet T)T + (v \bullet N)N. \tag{1.1}$$

The most familiar example of an orthonormal basis is the *standard basis* $T = (1, 0)$, $N = (0, 1)$. However, in Chapter 5 we will see that orthonormal bases can be associated in a very natural way to general points on curves, and that the way in which they change as we move along the curve provides geometric information of fundamental importance.

Exercises

1.3.1 Let u, v be any vectors in the plane. Establish the *parallelogram law*

$$|u + v|^2 + |u - v|^2 = 2|u|^2 + 2|v|^2.$$

1.3.2 The length of a vector was expressed in terms of the scalar product. Conversely, show that the scalar product can be expressed in terms of the length via the *Polarization Identity*

$$u \bullet v = \frac{1}{2}\left\{|u|^2 + |v|^2 - |u-v|^2\right\}.$$

1.4 The Complex Structure

One of the gains of working throughout with *planar* curves is that we will be able to take advantage of the fact that the points $z = (x, y)$ in the plane can be identified with complex numbers $z = x + iy$. Under this identification, vector addition of points in the plane corresponds to the usual addition of complex numbers. The gain is largely one of notational and computational efficiency (rather than mathematical understanding) but is still worthwhile. We will adopt standard notations by writing $x = \Re z$ for the *real part* and $y = \Im z$ for the *imaginary part* of the complex number $z = x + iy$. Likewise, the *complex conjugate* is written $\bar{z} = x - iy$, identified with the reflexion $z = (x, -y)$ of the point $z = (x, y)$ in the *x*-axis. The real gain lies in the availability of multiplication and division for complex numbers. Recall that the *product* of two complex numbers $z = x + iy$, $w = u + iv$ is defined to be

$$zw = (xu - yv) + i(xv + yu).$$

In particular, for any complex number $z = x + iy$, identified with the point $z = (x, y)$, we have $iz = -y + ix$ identified with the point $iz = (-y, x)$ obtained by rotating $z = (x, y)$ anticlockwise about the origin through a right angle.

Example 1.4 Recall that the component of a vector a in the direction of a unit vector b is the vector $(a \bullet b)b$. (Example 1.1.) It is useful to express this in complex notation. Note that for *any* two vectors a, b we have $2(a \bullet b) = a\bar{b} + \bar{a}b$: in particular when b is a *unit* vector (i.e. $b\bar{b} = |b|^2 = 1$) we have $2(a \bullet b)b = a + \bar{a}b^2$.

Note that the length of the vector $z = (x, y)$ is the *modulus* $|z|$ of the corresponding complex number $= x + iy$: in particular, vectors of unit length correspond to complex numbers of unit modulus. Recall that any vector $z = (x, y)$ of unit length can be written $z = (\cos t, \sin t)$ for some real number t: under the identification with complex numbers this means that any unit complex number u can be written $u = e^{it}$ where by definition $e^{it} = \cos t + i \sin t$. More generally, any complex number $z \neq 0$ can be written uniquely in the polar form $z = re^{it}$ where $r = |z|$.

1.5 Lines

Lines will play a fundamental role in studying the geometry of general curves, so it is worth recalling some basic facts. We define a *line* to be the set L of points (x, y) satisfying an equation of the form $ax + by + c = 0$, where a, b, c are real numbers, and at least one of a, b is non-zero. In this section we will review the basic properties of lines via a series of examples. The first example expresses the most basic property of all, namely that the equation of a line is determined (up to scalar multiples) by any two of its points.

Example 1.5 Through any two distinct points $p = (p_1, p_2)$, $q = (q_1, q_2)$ in $\mathbb{R}^2$ there is a unique line $ax + by + c = 0$. We seek scalars a, b, c (not all zero) for which

$$ap_1 + bp_2 + c = 0, \qquad aq_1 + bq_2 + c = 0.$$

Since p, q are distinct, the 2×3 coefficient matrix of these two linear equations in a, b, c has rank 2. By linear algebra it has kernel rank 1, so there is a non-trivial solution (a, b, c), and any other solution is a non-zero scalar multiple of this one. Explicitly, the line joining p, q has the equation

$$(p_1 - q_1)(y - p_2) = (p_2 - q_2)(x - p_1).$$

What is important for our purposes is that lines can be 'parametrized' in a natural way, providing a model for the general 'parametrized' curves of Chapter 2.

Example 1.6 Consider a line $ax + by + c$, and *distinct* points $p = (p_1, p_2)$, $q = (q_1, q_2)$ on the line. Then a brief calculation verifies that for any choice of scalar t the point $r = (1 - t)p + tq$ also lies on the line. Conversely, we claim that any point $r = (r_1, r_2)$ on the line has the form $r = (1 - t)p + tq$ for some scalar t. The proof goes as follows. Since p, q, r all lie on the line we have

$$\begin{cases} ap_1 + bp_2 + c = 0 \\ aq_1 + bq_2 + c = 0 \\ ar_1 + br_2 + c = 0. \end{cases}$$

That is a linear system of three equations in a, b, c. Since at least one of a, b is non-zero, the system has a non-trivial solution. By linear algebra, the 3×3 matrix of coefficients is singular, so the rows $(p_1, p_2, 1)$, $(q_1, q_2, 1)$, $(r_1, r_2, 1)$ are linearly dependent. However, the first two rows are linearly independent (as p, q are distinct) so the third row is a linear combination of the first two; thus there exist scalars s, t for which

$$(r_1, r_2, 1) = s(p_1, p_2, 1) + t(q_1, q_2, 1).$$

It follows immediately that $r = sp + tq$ and $1 = s + t$ from which we deduce the required relation $r = (1-t)p + tq$.

In view of this result we introduce the following definitions. Given two distinct points $p = (p_1, p_2)$, $q = (q_1, q_2)$ the *standard parametrized* line through p, q is the specific parametrization given by the formulas below, with the points p, q corresponding respectively to the parameters $t = 0$, $t = 1$.

$$x = (1-t)p_1 + tq_1, \qquad y = (1-t)p_2 + tq_2.$$

Example 1.7 The standard parametrization of a line depends on the choice of points p, q. For instance we can parametrize the line $y = 0$ via the points $p = (0, 0)$, $q = (1, 0)$ to obtain the parametrization $x = t$, $y = 0$: on the other hand the points $p = (-1, 0)$, $q = (1, 0)$ give rise to the parametrization $x = 2t - 1$, $y = 0$.

Example 1.8 The scalar product provides a convenient way of writing down the equation of a line. For any fixed non-zero vector $N = (a, b)$ the vectors $z = (x, y)$ orthogonal to N are those for which $ax + by = 0$. That is a single linear equation in the variables x, y satisfied by $x = -b$, $y = a$, so by linear algebra its solutions (x, y) lie on a line through the origin spanned by the vector $(-b, a)$. Thus the equation $ax + by = 0$ of any line through the origin can be written in the form $N \bullet z = 0$ for some non-zero vector N. More generally, the equation $ax + by + c = 0$ of *any* line can be written in the form $N \bullet z + c = 0$ for some non-zero vector N. Since the equation of a line is unique up to scalar multiples, the vector N likewise is unique up to scalar multiples: it is an example of a 'normal' vector. (Chapter 2.) In particular, in writing down the equation of a line we can always choose N to be a *unit* vector.

Example 1.9 Let N be a non-zero vector. Given a fixed point z_0, the equation of the line L through z_0 orthogonal to N can be written in the

1.5 Lines

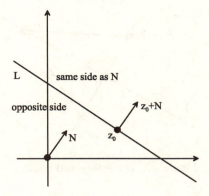

Figure 1.2. The two sides of a line

form $(z - z_0) \bullet N = 0$: indeed (as we saw in the previous example) that is the equation of a line orthogonal to N, and it certainly passes through z_0. For visualization purposes it helps to think of N with its origin at the point z_0: more precisely, we think in terms of $z_0 + N$ rather than N. Observe that the plane is partitioned into three sets by the relations

$$(z - z_0) \bullet N > 0, \quad (z - z_0) \bullet N = 0, \quad (z - z_0) \bullet N < 0$$

representing respectively one side of L, the line L itself and the other side of L. At a later stage in this book it will help us to be clear about which side is which. To this end, write θ for the angle between the vectors $z - z_0$ and N, so that

$$(z - z_0) \bullet N = \cos\theta |z - z_0||N|.$$

It follows immediately that the sign of $(z - z_0) \bullet N$ agrees with that of $\cos\theta$. Looking at the graph of the cosine function we see that the sign is positive if and only if $-\pi/2 < \theta < \pi/2$. One vector z on that side of L is the vector $z = z_0 + N$: for that reason we refer to the side of L with $(z - z_0) \bullet N > 0$ as being on the *same side* as N, and the other side as the *opposite side*.

Example 1.10 Let $N = (a, b)$ be a non-zero vector, and let L be a line with equation $z \bullet N = c$. By a *direction* for the line we mean any vector T orthogonal to N: it is an example of a 'tangent' vector. (Chapter 2.) Thus we could take $T = (-b, a)$. Alternatively, we could choose any two distinct points p, q on the line, and take $T = q - p$: the relation

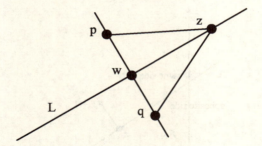

Figure 1.3. The orthogonal bisector

$T \bullet N = 0$ then follows from $p \bullet N + c = 0$, $q \bullet N + c = 0$ on subtraction. On this basis we say that two lines $a_1 x + b_2 y + c_1 = 0$, $a_2 x + b_2 y + c_2 = 0$ are *orthogonal* when the corresponding 'normal' vectors $N_1 = (a_1, b_1)$, $N_2 = (a_2, b_2)$ are orthogonal, or equivalently the corresponding 'tangent' vectors $T_1 = (-b_1, a_1)$, $T_2 = (-b_2, a_2)$ are orthogonal: either way, the condition is that $a_1 a_2 + b_1 b_2 = 0$.

Example 1.11 Let p, q be distinct points. A point z is *equidistant* from p, q when the distance from z to p equals the distance from z to q: the set of all points equidistant from p, q is called the *orthogonal bisector* of the line segment joining p, q. (Figure 1.3.) The orthogonal bisector is a line. Indeed the constraint on z is that $|p - z|^2 = |q - z|^2$: expanding both sides of this relation we obtain

$$2(q - p) \bullet z = |q|^2 - |p|^2$$

which is the equation of a line L, orthogonal to the vector $(q - p)$. Note that the *mid-point* $w = \frac{1}{2}(p + q)$ of the line segment joining p, q lies on the orthogonal bisector. (A formal definition of the term 'line segment' will be given in Chapter 2.)

Example 1.12 Using complex notation, consider two points $z = x + iy$, $w = u + iv$ equidistant from 0, and from 1. We claim that either $z = w$ or $z = \bar{w}$. The equidistance conditions are expressed by the relations

$$x^2 + y^2 = u^2 + v^2, \quad (x - 1)^2 + y^2 = (u - 1)^2 + v^2.$$

Subtracting these relation yields $u = x$, and hence $v = \pm y$: in the '+' case $w = z$, and in the '−' case $w = \bar{z}$.

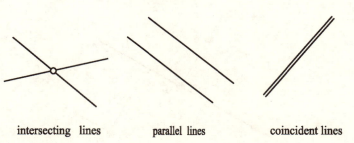

intersecting lines   parallel lines   coincident lines

Figure 1.4. Three ways in which lines can intersect

Example 1.13 To find the intersection of two lines $ax + by + c = 0$, $a'x + b'y + c' = 0$ we think of these as two linear equations in x, y and appeal to elementary linear algebra. When $\delta = ab' - a'b \neq 0$ the equations have a unique solution, and the intersection comprises a unique point: otherwise there are no solutions (*parallel lines*) or a line of solutions (*coincident lines*). (Figure 1.4.)

Exercises

1.5.1   Find the equation of the line through the points $p = (1, 1)$, $q = (4, 3)$, and the equation of the orthogonal line through $r = (-1, 1)$.

1.6 Projection to Lines

One of the key Euclidean constructions in this book will be the 'orthogonal projection' of a point onto a line. The situation is best summed up in one result.

Lemma 1.3 *Let L be a line and let p be a point. There is a unique point w on L for which $(w - p)$ is orthogonal to L. Moreover, w is given by the following formulas:*

$$w = u + \{(p - u) \bullet T\}T = p - \{(p - u) \bullet N\}N,$$

where u is any point on L, T is a unit vector in the direction of L, and N is a unit vector orthogonal to T.

Proof L can be parametrized as $z(t) = u + tT$. We seek a point $z(t)$ for which $z(t) - p$ is orthogonal to the direction T of the line, i.e. for which

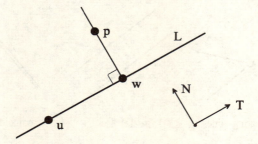

Figure 1.5. Orthogonal projection onto a line

$0 = (p - z(t)) \bullet T = (p - u - tT) \bullet T$. This relation has a unique solution $t = (p - u) \bullet T$. Substituting for t in $z(t)$ gives the first formula for w in the statement of the result. The second formula is immediate from the relation (1.1) with $v = p - u$. □

The point w appearing in Lemma 1.3 is called the *orthogonal projection* of p onto the line L, and the distance $|w - p|$ is the *distance* from the point p to the line L. (Figure 1.5.) The reason for giving two formulas for w is that sometimes one may prove more convenient than the other.

Example 1.14 Consider any 'vertical' line L with equation $x = x_0$, and any point $p = (x, y)$. The orthogonal projection w of p onto L is very easily determined. We could choose $u = (x_0, 0)$, $T = (0, 1)$, $N = (1, 0)$: with these choices $(p - u) \bullet T = y$, $(p - u) \bullet N = x - x_0$ and

$$w = u + \{(p - u) \bullet T\}T = p - \{(p - u) \bullet N\}N = (x_0, y)$$

as one might expect. The distance from p to L is $|w - p| = |x_0 - x|$.

Exercises

1.6.1 Let L be a line and let p be a point. Show that the orthogonal projection of p onto L is the unique point w on L for which the distance from p to w is minimized.

1.6.2 Let L be the line $ax + by + c = 0$ and let $p = (\alpha, \beta)$ be a point. Show that the distance d from p to L is given by the formula

$$d = \frac{|a\alpha + b\beta + c|}{a^2 + b^2}.$$

2
Parametrized Curves

Plane parametrized curves arise naturally throughout the physical sciences and mathematics. In this introductory chapter we will set up the underlying definitions, and introduce the reader to a small zoo of curves to provide a useful basis for illustration in later chapters. The first step in our geometric development is to associate 'tangent' vectors to each parameter on a curve. In particular, that allows us to distinguish 'irregular' parameters for which the associated 'tangent' vector is zero. Such parameters may correspond to points where the curve is visibly different from other points.

2.1 The General Concept

For the purposes of this book a *parametrized curve* (or just *curve* if there is no ambiguity) is a smooth mapping $z : I \to \mathbb{R}^2$, with $I \subseteq \mathbb{R}$ an open interval. Thus I is a set of real numbers t (the *parameters*) which satisfy an inequality of the form $a < t < b$, where we allow either or both of the possibilities $a = -\infty$, $b = \infty$.

The meaning of the term 'smooth' in the above is as follows. For each parameter t we have a point $z(t) = (x(t), y(t))$ in the plane: the resulting functions $x, y : I \to \mathbb{R}$ of a single real variable t are the *components* of z, and 'smoothness' means that at every parameter t both components x, y possess derivatives of all orders. In all the examples which arise in this book it will be self evident from elementary analysis that z is indeed 'smooth' in this sense. The term 'smooth', though excluding gross pathologies, does still allow unexpected phenomena. For instance, it is shown in books on analysis that there exist *continuous* mappings $z : I \to \mathbb{R}^2$, with I an open interval, for which the trace is the whole of $\mathbb{R}^2$ (the so-called *Peano curves*). The condition that z is smooth eliminates

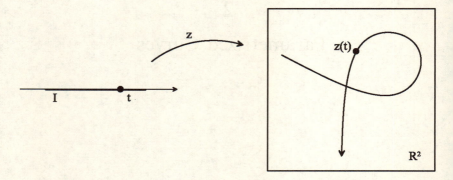

Figure 2.1. A parametrized curve

this type of behaviour. However, it does not preclude the possibility of the trace having sharp 'corners'. For most of the curves in this book the domain I will be the whole real line $\mathbb{R}$: only infrequently, when the domain is not $\mathbb{R}$, will we specify it. Sometimes there will be practical advantages in thinking of $z(t)$ as a complex number, in which case we may change our notation and write $z(t) = x(t) + iy(t)$.

Example 2.1 According to Example 1.6 the line through any two distinct points $p = (p_1, p_2)$, $q = (q_1, q_2)$ is naturally parametrized as

$$x(t) = (1-t)p_1 + tq_1, \quad y(t) = (1-t)p_2 + tq_2$$

with the parameters $t = 0$, $t = 1$ corresponding to the points p, q. Using complex notation this can be written more concisely as

$$z(t) = (1-t)p + tq.$$

Example 2.2 The *circle* of radius r with centre $z_0 = (x_0, y_0)$ is the set of points $z = (x, y)$ for which the distance between z, z_0 is the fixed positive real number r: put another way, a circle is the set of points (x, y) satisfying an equation of the form

$$(x - x_0)^2 + (y - y_0)^2 = r^2.$$

Two circles with the same centre (but possibly different radii) are *concentric*. The circle is naturally parametrized as

$$x(t) = x_0 + r\cos t, \quad y(t) = y_0 + r\sin t.$$

In complex number notation this would read instead $z(t) = z_0 + re^{it}$.

2.1 The General Concept

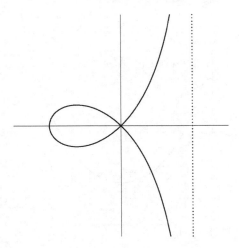

Figure 2.2. The right strophoid

A special case is the *unit circle* of radius 1 and centre the origin, with the parametrization $x(t) = \cos t$, $y(t) = \sin t$ or $z(t) = e^{it}$.

Example 2.3 A curve of historical interest is the *right strophoid*, the curve z with components x, y defined by the following formulas, where $a > 0$

$$x(t) = 2a\left(\frac{1-t^2}{1+t^2}\right), \quad y(t) = -2at\left(\frac{1-t^2}{1+t^2}\right).$$

The right strophoid illustrates a general feature, namely that its trace crosses itself at the origin: more precisely, there are two values $t = 1$, $t = -1$ of the parameter for which $z(t) = 0$. (Figure 2.2.) Later in this chapter we will formalize the idea of a 'self crossing' for a curve. Note that $x(t) \to 2a$ as $t \to \pm\infty$, so the line $x = 2a$ is an 'asymptote' of the curve, indicated by the vertical dotted line.

The *trace* of the curve $z : I \to \mathbb{R}^2$ is the set of points $z(t)$ with $t \in I$. The reader is encouraged to maintain a crystal clear mental distinction between the curve z and its trace. For instance the curves defined by $x(t) = t$, $y(t) = 0$ and $x(t) = 2t$, $y(t) = 0$ are different mappings, but have the same trace, namely the x-axis. Thus for a *constant* curve (one for which $z(t) = c$ for all t, with c a fixed point) the trace is a single point. A set of points Z in the plane is *parametrized* by a curve z when Z coincides with the trace of z.

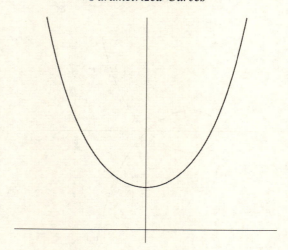

Figure 2.3. The catenary

Example 2.4 Let $f : I \to \mathbb{R}$ be any smooth function. The graph of f is defined to be the set of all points $(t, f(t))$ with $t \in I$, so is the trace of the curve $z : I \to \mathbb{R}^2$ defined by $z(t) = (t, f(t))$. The reader is probably familiar with the exercise of sketching such curves, at least in the case when f is a polynomial, using the techniques of elementary algebra (to find the points where the curve cuts the x-axis by solving the equation $f(x) = 0$) and elementary calculus (to find the local extrema and inflexions). An example is the function $f(x) = \cosh x$ for which the associated curve $z(t) = (t, \cosh t)$ is known as the *catenary*. (Figure 2.3.)

The catenary is of historical interest, representing the form adopted by a perfect inextensible chain of uniform density hanging from two supports. It was studied first by Galileo (who mistook it for a parabola) and later by Jacques Bernoulli the Elder (who obtained its true form). It is also of contemporary mathematical interest, being a plane section of the minimal surface (a soap film catenoid) spanning two circular discs, the only minimal surface of revolution.

Example 2.5 The restriction of a curve $z : I \to \mathbb{R}^2$ to any open subinterval of I is again a curve, known as an (open) *arc* of z. The subinterval is itself given by an inequality of the form $c < t < d$, where possibly $c = -\infty$ or $d = \infty$. The trace of the arc is a *curve segment*, and (provided c, d are finite) is said to *join* the points $z(c), z(d)$. For instance, given two distinct

2.1 The General Concept

points $p = (p_1, p_2)$, $q = (q_1, q_2)$ we saw in Example 2.1 that the standard parametrized line through p, q is the specific parametrization given by

$$x = (1-t)p_1 + tq_1, \quad y = (1-t)p_2 + tq_2$$

with p, q corresponding to the parameters $t = 0, 1$. The restriction of this parametrization to the interval $0 < t < 1$ is an arc of the line, whose trace is the *line segment* joining p, q. The trace of a restriction to an interval of the form $c < t < \infty$ or $-\infty < t < d$ is a *half line*.

The idea of a 'periodic' function extends naturally to curves. A curve $z : \mathbb{R} \to \mathbb{R}^2$ is *periodic*, when there exists a non-zero real number p (called a *period*) with the property that $z(t+p) = z(t)$ for all t. As in the case of functions, all periods of a non-constant periodic curve are integer multiples of a minimal positive period p, simply referred to as the *period* of the curve. Thus the natural parametrization of the circle in the previous example is periodic, with period 2π. However, there are other ways of parametrizing the circle. One approach is to consider the intersections of the circle with the lines through some fixed point p on the circle. (It is traditional to refer to the *pencil* of all lines through p.)

Example 2.6 Take for instance the circle of radius $|a|$ centred at the point $(a, 0)$ where $a \neq 0$. The (non-vertical) lines through $p = (0,0)$ have the form $y = tx$ for some scalar t. Substituting in the equation $x^2 + y^2 = 2ax$ of the circle we see that such a line meets the circle at a further point (Figure 2.4) with coordinates

$$x = \frac{2a}{1+t^2}, \quad y = \frac{2at}{1+t^2}.$$

That gives a second parametrization, *not of the whole circle, but of the circle with the point p deleted*. There is nothing special about the choice of the point p on the circle. For instance, choosing instead the pencil of lines $y = t(2a - x)$ through the point $p = (2a, 0)$ we obtain another parametrization

$$x = \frac{2at^2}{1+t^2}, \quad y = \frac{2at}{1+t^2}.$$

In Chapter 4 we will clarify the relation between different parametrizations of the same curve via the concept of 'parametric equivalence'.

Example 2.7 An attractive family of curves is provided by the *rose curves* (or *rhodonea*) given by $z(t) = 2be^{it} \cos nt$ where b, n are non-zero

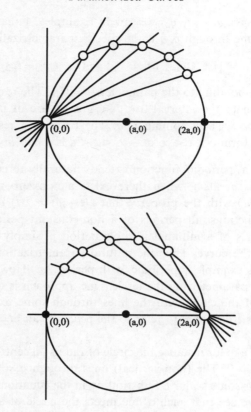

Figure 2.4. Parametrizing the circle

real numbers. Clearly, it is no restriction to assume that n is positive. Note that $|z(t)| = 2b|\cos nt| \le 2b$, so the trace of the curve lies inside the circle of radius $2b$ centred at the origin. The form of the curve depends on n. Our main interest will be the case when n is a positive *integer*, when the curves are known as *clover leaves*. In that case the curve touches the circle exactly when $\cos nt = \pm 1$, $\sin nt = 0$, i.e. when e^{it} is a $2n^{\text{th}}$ complex root of 1. Note that z has period 2π when n is even, but period π when n is odd: thus when n is even we obtain exactly $2n$ points of contact (the *$2n$-leaved clover*) and when n is odd exactly n (the *n-leaved clover*). An exceptional case arises when $n = 1$, and the trace is the circle of radius b centred at the point $(b, 0)$. (Exercise 2.1.2.) For $n = 2, 3, \ldots$ we obtain the series of curves whose first few members are illustrated in Figure 2.5.

2.1 The General Concept

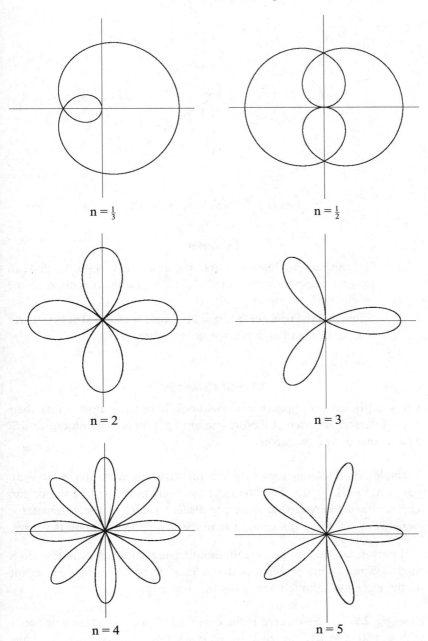

Figure 2.5. Some rose curves

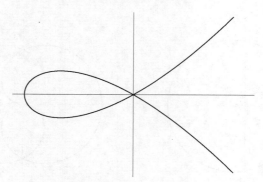

Figure 2.6. Tschirnhausen's cubic

Exercises

2.1.1 By considering lines through the point $p = (-a, 0)$, find an explicit parametrization of the circle of radius a centred at the origin, with the point p deleted.

2.1.2 Show that the rose curve $z(t) = 2be^{it} \cos t$ with $b > 0$ parametrizes the circle of radius b centred at the point $(b, 0)$.

2.2 Self Crossings

Many of the curves appearing in this book have the property that their traces intersect themselves. Before presenting a formal definition we will look at one or two examples.

Example 2.8 *Tschirnhausen's cubic* is the curve given by $x(t) = 3(t^2 - 3)$, $y(t) = t(t^2 - 3)$. Figure 2.6 shows the curve crossing itself at the origin. The key factor here is that there are *distinct* values of the parameter t, namely $t = \sqrt{3}$, $t = -\sqrt{3}$, giving rise to the *same* point $(0, 0)$ on the curve.

However, we have to be careful about basing a formal definition on a single example. One problem is that z may be periodic, so *every* point on the trace has infinitely many distinct preimages.

Example 2.9 The *eight-curve* is the curve $x(t) = a \cos t$, $y(t) = a \sin t \cos t$ with $a > 0$. Clearly, the eight-curve is periodic, with period 2π. Thus for *any* choice of t there are infinitely many distinct values of the parameter, namely $t + 2n\pi$ with n an integer, which map to the same

2.2 Self Crossings

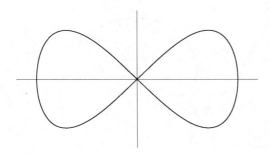

Figure 2.7. The eight-curve

point on the trace. However Figure 2.7 suggests that the origin should be a 'self crossing'. The property that differentiates the origin from every other point on the trace is that there exist distinct parameters $s = \pi/2$, $t = 3\pi/2$ for which $z(s) = z(t) = (0,0)$, but the difference $p = \pi$ is *not* a period of the curve.

On this basis we define a *self crossing* of a curve z to be a point on the trace for which there exist distinct values s, t of the parameter with $z(s) = z(t)$, and such that $p = s - t$ is not a period. With this definition the origin is a self crossing of the eight-curve.

Example 2.10 To find the self crossings of the curve $x(t) = t^2 + t^3$, $y(t) = t^3 + t^4$ we require distinct parameters s, t with $x(s) = x(t)$, $y(s) = y(t)$, i.e. $s^2(1+s) = t^2(1+t)$, $s^3(1+s) = t^3(1+t)$, yielding solutions $s = -1$, $t = 0$ with $s < t$. We conclude that $(0,0)$ is a self crossing, indeed the only self crossing on the curve.

Example 2.11 The curve $x(t) = \cos^3 t \cos 3t$, $y(t) = \cos^3 t \sin 3t$ illustrated in Figure 2.8 is *Cayley's sextic*. Clearly, it is periodic, with period π. The illustration suggests that the curve has a self crossing. Indeed, that is the case: the parameters $s = 2\pi/3$, $t = \pi/3$ produce a self crossing at the point $(-1/8, 0)$ on the trace.

Example 2.12 The curve $x(t) = \sin t$, $y(t) = 0$ is periodic, with period 2π, having trace the interval of the x-axis given by $-1 \le x \le 1$. For this example *every* point on the trace with $-1 < x < 1$ is a self crossing: indeed for such an x a glance at the graph of the sine function shows there exist distinct parameters s, t with $|s - t| < 2\pi$ for which $x = \sin s = \sin t$.

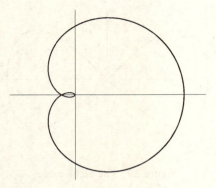

Figure 2.8. Cayley's sextic

Example 2.13 Figure 2.5 suggests that for $n \geq 2$ the origin is a self crossing of the clover leaf $z(t) = 2be^{it}\cos nt$, where $b > 0$, and n is a positive integer. Recall that the clover leaves are periodic, with period π when n is odd, and period 2π when n is even. (Example 2.7.) For self crossings we seek distinct parameters s, t differing by less than the period for which $z(s) = z(t)$, i.e. $e^{is}\cos ns = e^{it}\cos nt$. Assume both sides of this relation are non-zero. Taking moduli gives $\cos ns = \pm\cos nt$, so $e^{is} = \pm e^{it}$. In the '+' case s, t differ by a multiple of 2π so we cannot find s, t satisfying our requirements. In the '−' case s, t differ by a multiple of π, so we can suppose n is even: but in that case $\cos ns = \cos nt$ leading to $\cos ns = 0$, $\cos nt = 0$ contradicting our assumption. It remains to consider the case when $\cos ns = \cos nt = 0$. When n is odd, any distinct choices of s, t from the list of n values

$$\frac{\pi}{2n}, \frac{3\pi}{2n}, \ldots, \frac{(2n-1)\pi}{2n}$$

give rise to a self crossing at the origin: and when n is even we can take any distinct choices of s, t from the list of $2n$ values

$$\frac{\pi}{2n}, \frac{3\pi}{2n}, \ldots, \frac{(4n-1)\pi}{2n}.$$

Exercises

2.2.1 In each of the following cases sketch the graphs of the individual functions $x(t)$, $y(t)$, sketch the curve $z(t) = (x(t), y(t))$, and find

2.3 Tangent and Normal Vectors

any self crossings.

(i) $x(t) = 1 + t^2$, $y(t) = t^3$
(ii) $x(t) = 1 + t^2$, $y(t) = t(t^2 + 1)$
(iii) $x(t) = t^2$, $y(t) = t^3 + t^4$
(iv) $x(t) = t^4$, $y(t) = t + t^2$
(v) $x(t) = t^2 + t^3$, $y(t) = t^4$
(vi) $x(t) = t^2$, $y(t) = t^5 - t^3$.

2.2.2   Show that the self crossing of the Cayley sextic described in Example 2.11 is unique.

2.2.3   A curve is defined by the formula below. Show that the curve has exactly one self crossing.

$$x(t) = \frac{-t^3}{1 + t^4}, \quad y(t) = \frac{t}{1 + t^4}.$$

2.2.4   *Maclaurin's trisectrix* is the curve given by the formula below. Show that the curve has exactly one self crossing.

$$x(t) = \frac{t^2 - 3}{1 + t^2}, \quad y(t) = \frac{t(t^2 - 3)}{1 + t^2}.$$

2.3 Tangent and Normal Vectors

Given a curve z we define the *tangent vector* at the parameter t to be the vector $z'(t) = (x'(t), y'(t))$, where the dashes denote differentiation with respect to t. And the *normal vector* at t is the vector $iz'(t) = (-y'(t), x'(t))$ obtained by rotating the tangent vector through a right angle anticlockwise. One visualizes the tangent vector as in Figure 2.9 with its origin at the point $z(t)$, and thinks of it as representing the instantaneous direction of travel at t: likewise, one visualizes the normal

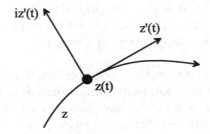

Figure 2.9. Tangent and normal vectors

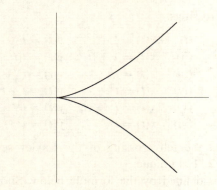

Figure 2.10. The semicubical parabola

vector with its origin at $z(t)$ orthogonal to the instantaneous direction of travel at t.

We say that t is a *regular* parameter of the curve z when $z'(t) \neq 0$: otherwise it is *irregular*. A curve z is *regular* when every parameter is regular. We expect a general parameter t to be regular, and at most a discrete set of parameters to be irregular.

Example 2.14 The rose curves $z(t) = 2be^{it}\cos nt$ of Example 2.7 with $b \neq 0$, $n \neq 0$ are regular. The tangent vector can be written in the form

$$z'(t) = bie^{it}\left\{(1+n)e^{nit} + (1-n)e^{-nit}\right\}$$

and is zero if and only if $(n+1)e^{2nit} = (n-1)$. Taking moduli, we deduce that $|n+1| = |n-1|$, contradicting the assumption $n \neq 0$.

Example 2.15 The *semicubical parabola* is the curve $x(t) = t^2$, $y(t) = t^3$. The tangent vector at t is $z'(t) = (2t, 3t^2)$, which vanishes if and only if $t = 0$. Thus $t = 0$ is the only irregular value of the parameter, corresponding to the 'cusp' point $(0,0)$ on the trace in Figure 2.10. We will have more to say about 'cusps' in Chapter 7.

Example 2.16 Let a, b be positive real numbers. Take C to be the circle radius a centred at the point $(a, 0)$, parametrized as in Example 2.2. And let L be the line $x = a^2/b$. For any point P on C, let Q be the point where the 'horizontal' line through P meets L, and let R be the point where the 'vertical' line through P meets the line joining Q to the origin. (Figure 2.11.) The locus of the point R is known as the *piriform*. A minor

2.3 Tangent and Normal Vectors

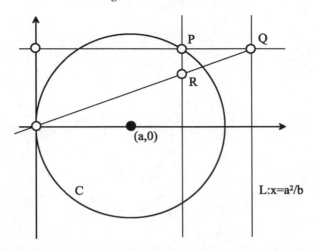

Figure 2.11. Construction of the piriform

calculation now shows that the piriform can be parametrized as

$$x(t) = a(1 + \cos t), \quad y(t) = b \sin t \, (1 + \cos t).$$

Irregular parameters are given by $0 = x'(t) = a \sin t$ and $0 = y'(t) = b(\cos 2t + \cos t)$. These are only satisfied when $\sin t = 0$, $\cos t = -1$, so at infinitely many parameters $t = (2n + 1)\pi$, where n is an integer, corresponding to the 'cusp' point $(0, 0)$ on the trace.

Example 2.17 The reader should not assume (on the basis of the last two examples) that an irregular parameter will always give rise to a point on the trace visibly different from most other points. For instance the x-axis can be parametrized as $x(t) = t^3$, $y(t) = 0$ with tangent vector $(3t^2, 0)$, vanishing when $t = 0$; but the corresponding point $(0, 0)$ can be distinguished in no way from any other point on the trace.

Some classes of curves are actually *defined* by conditions on their tangent vectors. Here is an interesting class of curves arising in Computer Aided Design (CAD) where the concept of 'tangent vector' plays a crucial role. The motivation is as follows. One is given a plane 'curve', for instance part of an artist's visualization of an industrial product, and one seeks a useful mathematical model for this curve which can be handled on a computer. The underlying idea was developed in the late 1950's by two design engineers working for rival French car companies, namely

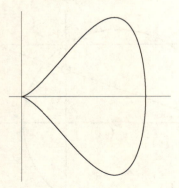

Figure 2.12. The piriform

Bézier (working for Renault) and de Casteljau (working for Citroën). A first crude step is to take a sequence of points $b_0, b_2, \ldots, b_{2n}$ on the curve and interpolate a polynomially parametrized curve. However, this process is intrinsically unsatisfactory: as n increases, so the degrees of the polynomials increase, and the interpolating curve may oscillate wildly. The idea is to control this oscillation by specifying the tangent direction at each point. One way of doing this is to associate to each point b_{2k} another point b_{2k+1} and stipulate that the tangent direction of the interpolating curve at b_{2k} should be the direction of the line joining b_{2k}, b_{2k+1}. The problem is then to write down an explicit curve with a given sequence of *control points* $b_0, b_1, \ldots, b_{2n}, b_{2n+1}$. (Figure 2.13.)

Example 2.18 Here is a solution for the case of four control points b_0, b_1, b_2, b_3. To these points we associate the *Bézier curve* defined by

$$z(t) = (1-t)^3 b_0 + 3t(1-t)^2 b_1 + 3t^2(1-t)b_3 + t^3 b_2.$$

The crucial property of this curve is that it passes through the points b_0, b_2 and has tangent directions $b_1 - b_0$, $b_2 - b_3$ at those points, since

$$z(0) = b_0, \quad z(1) = b_2, \quad z'(0) = 3(b_1 - b_0), \quad z'(1) = 3(b_2 - b_3).$$

Any Bézier curve has components $x(t)$, $y(t)$ both of which are polynomials of degree ≤ 3. Conversely, any such curve is a Bézier curve: one has only to observe that the polynomials $(1-t)^3$, $3t(1-t)^2$, $3t^2(1-t)$, t^3 are linearly independent, so form a basis for the space of polynomials of degree ≤ 3 in the variable t.

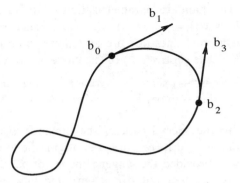

Figure 2.13. Curve with control points b_0, b_1, b_2, b_3

Exercises

2.3.1 Sketch the trace of the curve $x(t) = \cos^4 t$, $y(t) = \sin^4 t$. Find the irregular parameters, and the corresponding points on the trace.

2.3.2 In each of the following cases find the four control points for the given Bézier curves.

(a) The line $z(t) = (1-t)p + tq$ where p, q are distinct points.
(b) The parabola $x = at^2$, $y = 2at$ with $a > 0$.
(c) The curve $x(t) = t + t^2$, $y(t) = t^2 + t^3$.
(d) Tschirnhausen's cubic $x(t) = 3(t^2 - 3)$, $y(t) = t(t^2 - 3)$.

2.4 Tangent and Normal Lines

Suppose t is a regular parameter for the curve z. The *tangent line* to the curve at t is the line through $z(t)$ in the direction of the tangent vector $z'(t)$, and the *normal line* is the line through $z(t)$ in the direction of the normal vector $iz'(t)$. Bear in mind that tangent and normal lines are not defined at an irregular parameter t; their respective equations are

$$\begin{cases} (x - x(t))y'(t) - (y - y(t))x'(t) &= 0 \\ (x - x(t))x'(t) + (y - y(t))y'(t) &= 0. \end{cases}$$

Example 2.19 Consider the graph of a smooth function $y = f(x)$, regularly parametrized as $x(t) = t$, $y(t) = f(t)$. The tangent and normal lines at the parameter t have the respective equations

$$y = f(t) + f'(t)(x - t), \quad x = t + f'(t)(y - f(t)).$$

Example 2.20 The standard parametrization of the circle radius $r > 0$ with centre (a, b) is $x(t) = a + r\cos t$, $y(t) = b + r\sin t$. For this example the tangent and normal lines at t are as follows: note that the normal line passes through the centre (a, b) of the circle, as we would expect.

$$\begin{cases} x\cos t + y\sin t &= a\cos t + b\sin t + r \\ x\sin t - y\cos t &= a\sin t - b\cos t. \end{cases}$$

Example 2.21 In the special case of the previous example when the circle has unit radius, and its centre is the origin, the tangent line at t is $x\cos t + y\sin t = 1$. Provided the tangent line is not parallel to an axis, it intersects the x-axis at the point $X(t) = \sec t$, and the y-axis at the point $Y(t) = \csc t$, so the lines parallel to the axes through the intersections meet at the point $Z(t) = (X(t), Y(t))$. Eliminating t from these relations we see that $Z(t)$ lies in the set of points in the plane satisfying the relation

$$\{(X, Y) : X^2 + Y^2 = X^2 Y^2\}.$$

Figure 2.14 illustrates this set, known as the *cross curve*. It comprises four 'branches', each having two asymptotes, plus an isolated point at the origin. Each 'branch' is parametrized by the formula $Z(t)$ restricted to one of the four intervals obtained from the interval $0 < t < 2\pi$ by deleting the points $t = \pi/2$, $t = \pi$, $t = 3\pi/2$.

Example 2.22 The *tractrix* is the curve z having the components $x(t) = t - \tanh t$, $y(t) = \operatorname{sech} t$. The reader will readily verify that the tractrix has

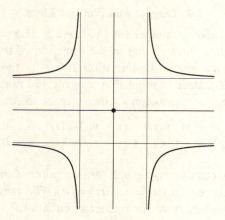

Figure 2.14. The cross curve

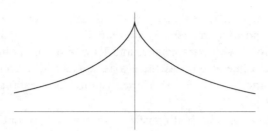

Figure 2.15. The tractrix

just one irregular parameter, namely $t = 0$. At any other parameter the tangent line has equation $x - (\sinh t)y = t$. The tractrix has the following remarkable property: the line segment joining the point $z(t)$ to the point $(t, 0)$ where the tangent line meets the x-axis, has constant unit length. (Figure 2.15.) There is another way of saying this. The circle of unit radius centred at the point $(t, 0)$ passes through the point $z(t)$, and the tangent line to the circle at $z(t)$ is orthogonal to the tangent line to the tractrix at that point. Thus the tractrix has the property that it meets all circles of unit radius centred on the x-axis orthogonally. For that reason the tractrix is described as an 'orthogonal trajectory' of that family of circles.

The tractrix gives rise to an interesting example in the elementary geometry of differentiable *surfaces*: the surface of revolution obtained by rotating it about the x-axis is the *pseudosphere*, distinguished by the surprising property of having constant negative Gaussian curvature.

Exercises

2.4.1 Find the tangent line at the parameter $t = 1$ to the curve defined by
$$x(t) = \frac{t^3 - 1}{3t^2 - 1}, \quad y(t) = \frac{t + 1}{3t^2 - 1}.$$

2.4.2 Find the tangent lines to the curve $x(t) = t$, $y(t) = t^4 - t + 3$ which pass through the origin.

2.4.3 Let P be the point where the tangent line to the curve $x(t) = t$, $y(t) = t^3$ at the parameter t meets the x-axis, and let $N = (t, 0)$. Show that $OT = 2TN$ where O is the origin. Generalize the result to the curve $x(t) = t$, $y(t) = t^n$.

2.4.4 The curve $x(t) = t^m$, $y(t) = t^{-n}$ is defined for $t > 0$, where m, n are positive integers. Show that the curve is regular. Let p be the point with parameter t, and let q, r be the points where the tangent line at p meets the x-axis, y-axis respectively. Show that the ratio $|p-q|/|p-r|$ of the distances is constant, and find its value.

2.4.5 Let k, a, b, c be real constants with $k \neq 0$. A smooth function $y = f(x)$ is defined by the following formula. Show that the tangent line at the point on the graph with $x = (a+b)/2$ passes through $(c, 0)$.

$$f(x) = k(x-a)(x-b)(x-c).$$

2.4.6 In the following f is the smooth function defined by $f(x) = x^2 - x^3$. For each x write $P(x) = (x, f(x))$ for the corresponding point on the graph $y = f(x)$.

(a) Find the coordinates of the point $Q(x)$ where the tangent line at $P(x)$ cuts the graph again.

(b) Let P_1, P_2, P_3 be three points on the graph, and let Q_1, Q_2, Q_3 be the points where the tangents at P_1, P_2, P_3 meet the graph again. Show that Q_1, Q_2, Q_3 are collinear if and only if P_1, P_2, P_3 are collinear.

(c) Show that the locus of the mid-point of the segment PQ is the graph of $g(x) = 1 - 9x + 28x^2 - 28x^3$.

2.4.7 A regular curve z has the property that all the tangent lines pass through a fixed point p. Show that the trace of z coincides with that of a line segment.

2.4.8 Show that all the normal lines to the curve defined by the formulas

$$x(t) = r(\cos t + t \sin t), \quad y(t) = r(\sin t - t \cos t)$$

with $r > 0$ are equidistant from the origin. (A formula for the distance from a point to a line is given in Exercise 1.6.2.)

2.4.9 Given any point P on the parabola $x(t) = at^2$, $y(t) = 2at$ with $a > 0$ there exists a unique circle C centred at the focus $F = (a, 0)$ and passing through P. Show that the normal line at P passes through an intersection of C with the x-axis. (That gives a practical construction for the normal line to a parabola using only ruler and compasses.)

Exercises

2.4.10 A regular curve z has the property that all the normal lines are parallel. Show that the trace of z is a line segment.

2.4.11 A regular curve z has the property that all the normal lines pass through a fixed point c. Show that the trace of z is a circle segment, with centre c.

3
Classes of Special Curves

Over the course of time a wide diversity of special curves has arisen in the physical sciences. There is simply not space in a monograph of this nature to discuss them all, though some are introduced in the examples and exercises to illustrate general ideas. However, some classes of special curves deserve separate mention, such as lines, the most basic class of all. In the next section we review the standard conics, equally basic to any sensible development, and in Section 3.2 pursue the idea of implicitly defined curves a little further. Trochoids, curves traced by a point carried by one circle rolling on another, comprise the subject matter of Section 3.3. One of their attractions is that examples crop up naturally in physical problems. They also provide an introduction to roulettes, curves traced by a point carried by a general curve rolling on another. Roulettes are not just an amusing construction: as we will see in Chapter 13, they play a key role in planar kinematics.

3.1 The Standard Conics

For the moment a general *conic* is a set of points defined by the vanishing of a polynomial in two variables x, y of degree 2, so a set defined by an equation of the form

$$ax^2 + 2hxy + by^2 + 2gx + 2fy + c = 0$$

where a, b, c, f, g, h are real numbers, and at least one of a, b, h is non-zero. We do not intend to pursue the study of general conics. Rather, we concentrate on a class of conics which arises from the following classical construction. One is given a line L (the *directrix*), a point F (the *focus*) not on L, and a variable point P subject to the constraint that its distance from F is proportional to its distance from L. Write O_P

3.1 The Standard Conics

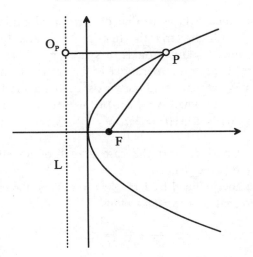

Figure 3.1. Parabola as a standard conic

for the orthogonal projection of P onto L. Then the constraint reads $PF = ePO_P$, for some positive constant of proportionality e, known as the *eccentricity*. Note that the line joining F to its orthogonal projection O_F on L is always an axis of symmetry, since reflexion in that line leaves the constraint unaltered. The locus of P is a *parabola* when $e = 1$, an *ellipse* when $e < 1$, or a *hyperbola* when $e > 1$. Figure 3.1 illustrates the construction for the case $e = 1$ of a parabola.

The 'standard conics' arise from the special case when L is the y-axis, $F = (k, 0)$ for some non-zero real number k, and $P = (X, Y)$, so $O_P = (0, Y)$: then, squaring both sides of the relation $PF = ePO_P$ we see that the coordinates X, Y satisfy the following relation, so the locus of P is indeed a conic:

$$(1 - e^2)X^2 - 2kX + Y^2 + k^2 = 0.$$

Note that circles cannot be constructed in this way: indeed, since the eccentricity is positive the coefficients of X^2, Y^2 in the displayed equation are *different*, whereas in the equation of a circle they are the *same*. Convenient forms for the equations of the standard conics can be obtained by translation parallel to the x-axis. Let us look at these in more detail.

Example 3.1 Consider first the case $e = 1$ of a parabola, and the translation of the plane defined by $X = x + \frac{1}{2}k$, $Y = y$. Then the equation of the conic reduces to that of a *standard* parabola $y^2 = 4ax$, where $a = k/2$ with directrix the line $x = -a$ and focus the point $F = (a, 0)$. The line $y = 0$ is an axis of symmetry of the parabola, and the point where it meets the parabola is the *vertex*. Consider now the family of lines $y = 2at$ parallel to the x-axis: each line meets the parabola just once, at the point where $x = at^2$. In this way we obtain a *standard parametrization* $x(t) = at^2$, $y(t) = 2at$.

Example 3.2 Consider next the case $e < 1$ of an ellipse, and the translation of the plane defined by $X = x + K$, $Y = y$ where K is the positive real number defined by $K(1 - e^2) = k$. Then the equation of the conic reduces to that of a *standard* ellipse

$$\frac{x^2}{a^2} + \frac{y^2}{b^2} = 1,$$

where the constants a, b are defined by $a = eK$, $b^2 = a^2(1 - e^2)$ and so satisfy $0 < b < a$. The lines $x = 0$, $y = 0$ are axes of symmetry of a standard ellipse. The points $(0, \pm b)$, $(\pm a, 0)$ where the axes meet the ellipse provide the four *vertices*. It is traditional to refer to a as the *major semiaxis* and b as the *minor semiaxis*. With these choices the directrix is the line L^- with equation $x = -a/e$, and the focus is the point $F^- = (-ae, 0)$. Note that the symmetry of the equation in x, y shows that there is a second directrix line L^+ with equation $x = a/e$ having a corresponding focus $F^+ = (ae, 0)$. The *centre* of a standard ellipse is the mid-point of the line segment joining the two foci, i.e. the origin. The circle having the same centre as the ellipse and passing through the points $(\pm a, 0)$ is known as the associated *auxiliary circle*.

Despite the fact that the circle does not appear as a standard conic, it is profitable to think of a circle (centred at the origin) as the limiting case of standard ellipses as $b \to a$: that corresponds to $e \to 0$, and the two foci F^-, F^+ coalescing at the centre of the circle.

Example 3.3 We can parametrize a standard ellipse as follows. Clearly, any point (x, y) satisfying the equation of a standard ellipse must be subject to the constraints $-a \le x \le a$, $-b \le y \le b$. A glance at the graph of the sine function shows that we can therefore write $y = b \sin t$ for some real number t. Substituting in the equation we see that $x = \pm a \cos t$. Choosing the '+' option we obtain a *standard parametrization* $x(t) = a \cos t$, $y(t) = b \sin t$ of the ellipse: it is regular, periodic of period 2π, and traces the ellipse anticlockwise. Thinking of the circle of radius a

3.1 The Standard Conics

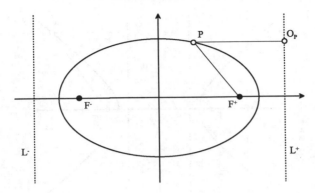

Figure 3.2. Ellipse as a standard conic

centred at the origin as the limiting case of the standard ellipse as $b \to a$ we recover a standard parametrization of the circle given in Example 2.2. The choice of the '−' option in the preceding analysis gives another parametrization of the ellipse, traversing the curve clockwise instead of anticlockwise.

Example 3.4 Finally, consider the case $e > 1$ of a hyperbola. Here we proceed almost exactly as we did for the ellipse. The same translation reduces the equation of the conic to that of a *standard* hyperbola

$$\frac{x^2}{a^2} - \frac{y^2}{b^2} = 1,$$

where the constants a, b are defined by $a = eK$, $b^2 = a^2(e^2 - 1)$. The lines $x = 0$, $y = 0$ are axes of symmetry of the hyperbola. Only the axis $y = 0$ meets the hyperbola, at the *vertices* $(\pm a, 0)$. Again we have directrix lines $x = -a/e$, $x = a/e$ with corresponding foci $F^- = (-ae, 0)$, $F^+ = (ae, 0)$. (Figure 3.3.) The *centre* of a standard hyperbola is the mid-point of the line segment joining the two foci, i.e. the origin. The lines $y = \pm bx/a$ are the *asymptotes* of the hyperbola: in texts on algebraic curves it is shown that they represent tangents to the curve at 'points at infinity'. The asymptotes are orthogonal if and only if $a = b$, in which case the hyperbola is said to be *rectangular*: that corresponds to the case when the eccentricity $e = \sqrt{2}$. A point (x, y) satisfying the equation of a standard hyperbola is subject only to the constraint that $x \geq a$ or $x \leq -a$: thus the key feature of a standard hyperbola is that it splits into two 'branches',

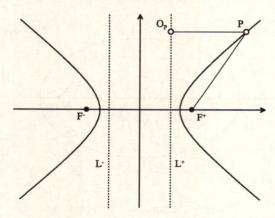

Figure 3.3. Hyperbola as a standard conic

namely the *positive branch* defined by $x \geq a$, and the *negative branch* defined by $x \leq -a$.

The parametrization of a standard hyperbola is more thought provoking than that of a standard ellipse. One approach is via the hyperbolic functions of elementary calculus.

Example 3.5 Let (x, y) be a point satisfying the equation of a standard hyperbola. Glancing at the graph of the sinh function we see that we can write $y = b \sinh t$ for a unique real number t. Then, substituting in the equation we see that $x = \pm a \cosh t$. The positive branch is then parametrized as $x(t) = a \cosh t$, $y(t) = b \sinh t$, and the negative branch by $x(t) = -a \cosh t$, $y(t) = b \sinh t$.

The choice of the hyperbolic functions here is by no means mandatory, indeed the only property of the sinh function we have used is that it is smooth and bijective. We could just as well take the tangent function of elementary trigonometry, which has the same property. Let us try this.

Example 3.6 For any point (x, y) on a standard hyperbola we can write $y = b \tan t$ for some real number t with $-\pi/2 < t < \pi/2$. Substituting for y in the equation of the hyperbola we obtain $x = \pm a \sec t$, leading to the parametrization $x(t) = a \sec t$, $y(t) = b \tan t$ for the positive branch, and $x(t) = -a \sec t$, $y(t) = b \tan t$ for the negative branch. At this point the perceptive reader might notice that the latter parametrization is obtained from the former by the simple device of replacing t by $t+\pi$. (The secant function changes sign, whilst the tangent function is left unchanged.) So one could think of the hyperbola as 'parametrized' by the single formula $x(t) = a\sec t$, $y(t) = b\tan t$: the positive branch corresponds to the interval $-\pi/2 < t < \pi/2$, and the negative branch to $\pi/2 < t < 3\pi/2$. However, that is *not* a parametrization in the strict sense of Chapter 2 because the domain is no longer an interval, but an interval from which a single point has been deleted.

This example illustrates a general situation. Quite often, natural geometric constructions give rise to mappings $z : D \to \mathbb{R}^2$ where the domain D is a union of several open intervals I, and the restrictions $z : I \to \mathbb{R}^2$ are smooth: the restrictions are then curves (in the strict sense) which can be studied by the methods developed in this book.

Exercises

3.1.1 Show that none of the parametrizations in Section 3.1 for the standard conics exhibit self crossings.

3.1.2 In each of the following cases show that the given curve satisfies the equation of a conic, and describe the trace.

(i) $x(t) = t^2$, $y(t) = t^4$
(ii) $x(t) = e^t$, $y(t) = e^{-t}$
(iii) $x(t) = \cos t$, $y(t) = \sin^2 t$
(iv) $x(t) = \cos^4 t$, $y(t) = \sin^4 t$.

3.1.3 Show that the tangent lines to the parabola $x(t) = at^2$, $y(t) = 2at$ at the parameters t_1, t_2 intersect at the point (X, Y) where $X = at_1 t_2$, $Y = a(t_1 + t_2)$.

3.1.4 Find those lines which are simultaneously tangent lines to the parabola $x(s) = s$, $y(s) = s^2$ at some parameter s, and normal lines to the parabola $x(t) = t$, $y(t) = -t^2/2$ at some parameter t.

3.1.5 Determine the minimum value of the distance PQ, where P and Q are the points where the tangent line to the ellipse $x(t) = a\cos t$, $y(t) = b\sin t$ meets the x- and y-axes. (It is assumed that the tangent line is parallel to neither axis.)

3.1.6 Determine the normals to the standard parametrized ellipse $x(t) = a\cos t$, $y(t) = b\sin t$ which are at a maximum distance from the centre, and calculate this distance. (A formula for the distance from a point to a line was given in Exercise 1.6.2.)

3.1.7 With the notation of Example 3.2 show that for any point P on a standard ellipse the following relation is satisfied. Conversely, show that any point P which satisfies this relation lies on a standard ellipse.

$$|P - F^+| + |P - F^-| = 2a.$$

3.1.8 With the notation of Example 3.4 show that for any point P on a standard hyperbola the following relation is satisfied. Conversely, show that any point P which satisfies this relation lies on a standard hyperbola.

$$||P - F^+| - |P - F^-|| = 2a.$$

3.1.9 Show that the tangent line at any parameter $t \neq 0$ to the rectangular hyperbola $x(t) = \pm \cosh t$, $y(t) = \sinh t$ intersects the x- and y-axes at points $\pm X(t)$, $Y(t)$ lying in the set known as the *bullet nose*, defined by

$$\{(X, Y) : Y^2 - X^2 = X^2 Y^2\}.$$

3.2 General Algebraic Curves

Many of the curves z which appear in this text can be studied profitably from the algebraic viewpoint. It is not our purpose here to pursue this viewpoint, developed in the companion volume 'Elementary Geometry of Algebraic Curves': rather, we wish to use the concept as a vehicle for further examples. Let us recall the basic definitions. An *algebraic curve* is a non-zero polynomial f given by a formula of the form

$$f(x, y) = \sum_{i,j} a_{ij} x^i y^j$$

where the sum is over finitely many pairs of non-negative integers i, j and the *coefficients* a_{ij} are real numbers. Two algebraic curves f, g are regarded as 'the same' when there is a non-zero scalar λ such that $g = \lambda f$. The *degree* of f is the maximal value of $i + j$ over the indices i, j with $a_{ij} \neq 0$, and in some sense measures the 'complexity' of the polynomial. The simplest algebraic curves are those of degrees 1, 2, 3, 4, ... known respectively as *lines*, *conics*, *cubics* and *quartics*. We met lines in Section 1.5 and conics in Section 1.3: numerous examples of cubics

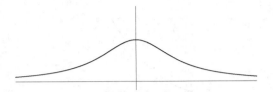

Figure 3.4. Agnesi's versiera

and quartics are scattered throughout the text. Here are two cubic curves of historical interest.

Example 3.7 The key idea of Example 2.6 is to use the pencil of lines through a point to parametrize a curve. Here is another illustration. Let $a > 0$, and consider the circle $x^2 + y^2 = 2ay$ of radius a centred at the point $(0, a)$. Each line $x = ty$ through the origin (Figure 3.4) meets the circle in a point P, and meets the line $y = 2a$ in a point Q. The 'horizontal' line through P then meets the 'vertical' line through Q in a point R, whose locus is the curve known as *Agnesi's versiera*. Substituting $x = ty$ in the equation of the circle we see that $R = (x(t), y(t))$ where

$$x(t) = 2at, \quad y(t) = \frac{2a}{1+t^2}.$$

The reader will readily check, by eliminating the variable t from these parametric equations, that the versiera satisfies the equation of the cubic curve $x^2 y = 4a^2(2a - y)$.

Example 3.8 According to Example 2.6 the lines $y = tx$ meet the circle of radius $a > 0$ centre $(a, 0)$ at the point

$$P = \left(\frac{2a}{1+t^2}, \frac{2at}{1+t^2} \right)$$

and the line $x = 2a$ at the point $Q = (2a, 2at)$. The *cissoid of Diocles* is the locus of the point R for which $R - O = Q - P$. (Figure 3.5.) Clearly R has the parametrization

$$x(t) = \frac{2at^2}{1+t^2}, \quad y(t) = \frac{2at^3}{1+t^2}.$$

Eliminating t from these relations we see that the cissoid satisfies the equation of the cubic curve $x^3 = (2a - x)y^2$.

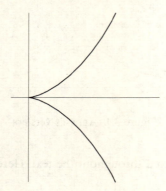

Figure 3.5. The cissoid of Diocles

The historical genesis of this curve lies in ancient Greek attempts to solve the problem of doubling the volume of a cube of given side a. They sought a geometric construction for the required side $2^{1/3}a$ of the doubled cube. The solution is due to Diocles in the second century BC, though another century passed before Geminus coined the description 'cissoid', meaning 'ivy shaped'. The solution is based on the observation that the line $y = 2(2a - x)$ joining the points $(2a, 0)$ and $(a, 2a)$ intersects the cissoid when $t = 2^{1/3}$: thus the line joining this intersection to the origin meets $x = a$ at the point with $y = 2^{1/3}a$.

Associated to any algebraic curve f is its *zero set*, the set of all points (x, y) for which $f(x, y) = 0$. One expects the zero set to be a 'curve' in some reasonable sense. However, be careful: the zero set of the conic $f(x, y) = x^2 + y^2 + c$ is infinite for $c < 0$, is a single point for $c = 0$, and is empty for $c > 0$. A parametrized curve z is said to be *algebraic* when there exists an algebraic curve f with the property that $f(x(t), y(t)) = 0$ identically in t where $x(t)$, $y(t)$ are the components of $z(t)$: put another way, the trace of z lies in the zero set of f. In that case there is no reason to suppose that the trace of z coincides with the zero set, a point illustrated by the next example.

Example 3.9 The curve $x(t) = a\cos^2 t$, $y(t) = b\sin^2 t$, where a, b are positive real numbers, satisfies the equation $bx + ay - ab = 0$ of the line joining the points $A = (a, 0)$, $B = (0, b)$. However, the components are subject to the conditions $0 \le x(t) \le a$, $0 \le y(t) \le b$ so the trace is the line segment joining A, B.

Exercises

Exercises

3.2.1 In each of the following cases show that the given curve is algebraic, and describe the trace.

(i) $x(t) = 2 + 3\sin t$, $y(t) = \sin t - 1$
(ii) $x(t) = 1 + e^t$, $y(t) = 5 + 2e^t$
(iii) $x(t) = a\cosh^2 t$, $y(t) = b\sinh^2 t$.

3.2.2 Find a function $f(x)$ with the property that its graph is the trace of Agnesi's versiera.

3.2.3 Show that there exists a cubic curve $f(x, y)$ such that every point on the parametrized curve $x(t) = \sin 2t$, $y(t) = \sin 2t \tan t$ satisfies $f(x(t), y(t)) = 0$.

3.2.4 Show that there exists a cubic curve $f(x, y)$ whose zero set contains the trace of the parametrized curve $x(t) = 1 + t^2$, $y = t + t^3$. Verify that there is exactly one point on the zero set of f which is not in the trace.

3.2.5 Descartes' folium is the cubic curve $x(x^2 + 3y^2) + (x^2 - y^2) = 0$. By considering the intersections of the curve with the pencil of lines $y = tx$ one obtains the following parametrization of the zero set. Show that the curve has exactly one self crossing.

$$x(t) = \frac{t^2 - 1}{3t^2 + 1}, \quad y(t) = \frac{t(t^2 - 1)}{3t^2 + 1}.$$

3.2.6 Let $a > 0$. The eight-curve is parametrized as $x(t) = a\cos t$, $y(t) = a\sin t \cos t$. (Example 2.9.) Show that there exists a quartic curve $f(x, y)$ such that $f(x(t), y(t)) = 0$ for all values of t. Let (x, y) be a point in the plane with $f(x, y) = 0$. Show that $-a \le x \le a$. Use this fact to show that there exists a real number t for which $x = x(t)$ and $y = y(t)$.

3.2.7 A parametrized curve is defined by $x(t) = t^2 + t^3$, $y(t) = t^3 + t^4$. Find a quartic curve $f(x, y)$ such that $f(x(t), y(t)) = 0$ for all t. (It helps to observe that $y = tx$.) Conversely, show that for any point (x, y) with $f(x, y) = 0$ there exists a real number t with $x = x(t)$, $y = y(t)$.

3.2.8 According to Example 2.16 the piriform is parametrized as

$$x(t) = a(1 + \cos t), \quad y(t) = b\sin t(1 + \cos t).$$

Show that there exists a quartic curve $f(x, y)$ such that $f(x(t), y(t)) = 0$ for all t. Conversely, show that for every point (x, y) with $f(x, y) = 0$ there exists a t for which $x = x(t)$, $y = y(t)$.

3.3 Trochoids

When one curve rolls (without slipping) along another fixed curve, any point which moves with the moving curve describes a curve, called a 'roulette'. (A French word meaning 'small wheel'.) The general concept will be studied in the context of planar kinematics in Chapter 13. However, in this section we will content ourselves with the special case when the curves are circles C, C' and the resulting roulette is called a *trochoid*. (A Greek word meaning 'wheel shaped'.) Incidentally, it is still possible to buy plastic *spirographs* which trace some of the more attractive trochoids, and were once considered suitable hobby material for older children: sadly, this kind of scientific hobby has almost disappeared in the face of overwhelming competition from electronic devices.

Trochoids occur naturally in the physical sciences, so it is worth writing down explicit parametrizations, and highlighting some of the cases. Think of C, C' lying in superimposed planes P, P' at time t_0: P is thought of as the *fixed plane* containing the *fixed circle* C, and P' as the *moving plane* containing the *moving circle* C'. The *centre* of the trochoid is defined to be the centre of the fixed circle C. A simple physical model can be made by taking P, P' to be plastic transparencies, and tracing C on P and C' on P' with differently coloured pens. The circle C' is then rolled along the circle C, carrying with it the plane P' until at time t we have the situation illustrated in Figure 3.6. In the moving plane we choose any

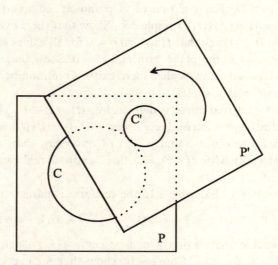

Figure 3.6. The idea of a roulette

3.3 Trochoids

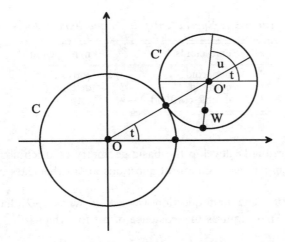

Figure 3.7. Construction of trochoids

tracing point W, fixed relative to the moving circle C'. Then, as P' moves, so W traces out a trochoid.

Assume that C has centre O at the origin, and radius $R > 0$, and that C' has centre O', and radius $R' \neq 0$. The case $R' > 0$ is interpreted as C' rolling on the outside of C (an *epitrochoid*), whilst $R' < 0$ is interpreted as C' rolling on the inside of C (a *hypotrochoid*). Suppose W is distant $R'h$ from the centre O' of the moving circle where $h > 0$. Write t for the angle between the line OO' and the x-axis, and assume (without loss of generality) that when $t = 0$ the point W lies on the x-axis.

It helps to use complex number notation. Write $z = z(t)$ for the position of W at 'time' t. Clearly $OO' = (R + R')e^{it}$. Now $OW = OO' - WO'$, so we have to compute WO'. Let u be the angle between the lines OO' and $O'W$. (Figure 3.7.) Then $WO' = hR'e^{i(t+u)}$, by elementary geometry. We can determine the angle u as follows. The arcs of C, C' along which moving has taken place have respective lengths Rt, $R'u$: since rolling takes place without slipping, these lengths are equal, and $u = Rt/R'$. Putting together the bits we obtain the parametrization

$$z(t) = (R + R')e^{it} - hR'e^{i(\frac{R+R'}{R'})t}. \tag{3.1}$$

The geometry of this curve depends crucially on the ratio $\lambda = R/R'$, and the scalar h. It will be no restriction to divide through by R', to obtain the parametrization in the more convenient form

$$z(t) = (\lambda + 1)e^{it} - he^{i(\lambda+1)t}. \tag{3.2}$$

Table 3.1. *Special epicycloids and hypocycloids*

λ	epicycloids	λ	hypocycloid
2	nephroid	$-5/2$	starfish
1	cardioid	-3	deltoid
1/2	double cardioid	-4	astroid

We will gradually develop the basic geometry of trochoids through a series of examples, asking natural questions about this class of curves.

Example 3.10 The first question we ask is when (3.2) has irregular parameters. That requires the existence of a t for which

$$z'(t) = i(\lambda + 1)e^{it}\{1 - he^{i\lambda t}\} = 0.$$

When $\lambda = -1$ *every* parameter is irregular: geometrically, the fixed and moving circles coincide, no rolling takes place, and the trace is a single point. Henceforth it will be tacitly assumed that $\lambda \neq -1$. With that assumption $z'(t) = 0$ if and only if $e^{i\lambda t} = 1/h$. It follows immediately that there exists an irregular parameter if and only if $h = 1$, i.e. *if and only if the tracing point lies on the circumference of the moving circle*. In Chapter 14 we will see that that is a special case of a general result in planar kinematics.

Example 3.11 In view of the previous example the case $h = 1$, when the tracing point P lies on the circumference of C', is of special significance. The form of the curve depends only on the ratio λ, and we can write the parametric equation as

$$z(t) = (\lambda + 1)e^{it} - e^{i(\lambda+1)t}.$$

For $\lambda > 0$ we have *epicycloids*, and for $\lambda < 0$ *hypocycloids*. Various names have been assigned traditionally to the curves arising by taking certain special values of λ: some of these are listed in Table 3.1, and illustrated in Figure 3.8. When $h = 1$ irregular parameters are given by $e^{i\lambda t} = 1$, so form the sequence

$$t = 0, \quad t = \pm\frac{2\pi}{\lambda}, \quad t = \pm\frac{4\pi}{\lambda}, \quad t = \pm\frac{6\pi}{\lambda}, \quad \ldots$$

The sequence of corresponding points on the trace is finite if and only if λ is rational. For instance when λ is an integer there are exactly $|\lambda|$

3.3 Trochoids

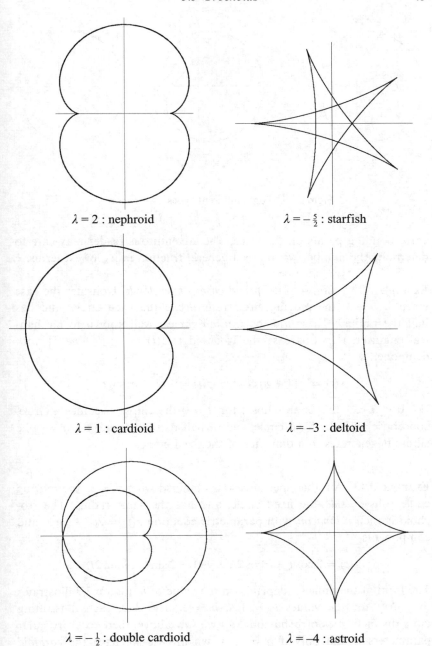

$\lambda = 2$: nephroid

$\lambda = -\frac{5}{2}$: starfish

$\lambda = 1$: cardioid

$\lambda = -3$: deltoid

$\lambda = -\frac{1}{2}$: double cardioid

$\lambda = -4$: astroid

Figure 3.8. Various epicycloids and hypocycloids

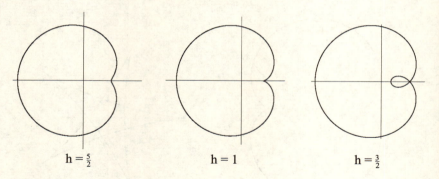

Figure 3.9. Three different types of limacon

corresponding points on the trace. The adventurous reader may care to determine the number when λ is a general fraction in its lowest terms.

Example 3.12 *Ellipses are special cases of trochoids.* Consider the case when $\lambda = -2$: the moving circle rolls *inside* the fixed circle, and has half the radius. (These are the *Cardan circles*, well known to mechanical engineers.) In this case the trochoid is $z(t) = -\{e^{it} + he^{-it}\}$ with components

$$x(t) = -(1+h)\cos t, \quad y(t) = -(1-h)\sin t.$$

For $0 < h < 1$ this is an ellipse: for $h = 0$ the ellipse becomes a circle, concentric with the fixed circle, and of half its radius: and for $h = 1$ the ellipse degenerates to a diameter of the fixed circle.

Example 3.13 Another special case is obtained when $\lambda = 1$: the moving circle rolls *outside* the fixed circle, and has the same radius. The trochoid is then a *limacon* with parametric equation $z(t) = 2e^{it} - he^{2it}$, and components

$$x(t) = 2\cos t - h\cos 2t, \quad y(t) = 2\sin t - h\sin 2t.$$

The form of the limacon depends on the value of h. Figure 3.9 illustrates the curve for the values $h = 3/4$, $h = 1$, $h = 3/2$, each illustrating quite distinct types of behaviour. As we saw above, there exist irregular parameters if and only if $h = 1$, in which case we have the *cardioid* (meaning 'heart shaped') providing a transitional case between $h < 1$ and $h > 1$.

3.3 Trochoids

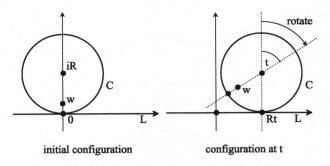

initial configuration   configuration at t

Figure 3.10. Construction of cycloids

Example 3.14 It is interesting to ask when self crossings appear on limacons. Note first that the limacon is periodic, with period 2π. Thus we require distinct values s, t of the parameter, whose difference is not a multiple of 2π, with $z(s) = z(t)$. The latter relation reduces to $e^{is} + e^{it} = 2/h$, where we can suppose $-\pi < s, t < \pi$: that can only happen when e^{is}, e^{it} are complex conjugate, i.e. $s = -t$ and $\cos s = 1/h$. Clearly, this relation can be solved for s in the given range if and only if $h > 1$. Limacons with $h > 1$ are said to be *nodal*. In Chapter 7 we shall discuss the case $h < 1$ in greater detail using another set of ideas.

Before leaving the subject of trochoids we should mention the limiting case which arises when you think of the fixed circle as being of 'infinite' radius, so a straight line L.

Example 3.15 Consider the roulette of a tracing point w carried by a circle C of radius $R > 0$ rolling along a straight line L. We will take L to be the x-axis. It is assumed that in the initial configuration C is the circle of radius R centred at the point iR, and that $w = iR(1 - h)$ is the point on the y-axis distance hR from the centre of C, where $h > 0$. (Figure 3.10.) Thus C is naturally parametrized as $q(t) = iR - iRe^{it}$, and L as $p(t) = Rt$. The reader will readily verify that the resulting roulette, known as a *cycloid*, has the parametrization

$$x(t) = R(t - h\sin t), \quad y(t) = R(1 - h\cos t). \tag{3.3}$$

The form of the cycloid depends on whether the tracing point is inside ($h < 1$), on ($h = 1$) or outside ($h > 1$) the moving circle. The three resulting forms are illustrated in Figure 3.11. For $h < 1$ we obtain a *curtate* (or 'shortened') cycloid reminiscent of the sine curve. When $h = 1$

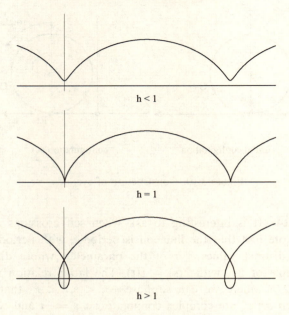

Figure 3.11. Forms of the cycloid

we obtain a *cuspidal* cycloid with infinitely many 'cusps' at the parameters $t = 2n\pi$, where n is an integer: the arc of the cycloid between any two consective 'cusps' is called an *arch*. Finally, for $h > 1$ we obtain a *prolate* (or 'extended') cycloid with infinitely many self crossings. (Exercise 3.3.5.)

Around the beginning of the eighteenth century Jacques Bernoulli the Elder (and others) discovered the *brachistochrone* property of the cycloid, namely that given two points A, B in a 'vertical' plane, the curve along which a particle takes the least time to slide from A to B is a cycloid. That observation represents the genesis of the area of mathematics now known as the theory of variations.

Exercises

3.3.1 Show that the astroid is (up to a scalar multiple) the curve $x(t) = \cos^3 t$, $y(t) = \sin^3 t$.

3.3.2 Find the tangent line at the point p with parameter t to the cardioid $x(t) = 2\cos t + \cos 2t$, $y(t) = 2\sin t - \sin 2t$. Show that the points q, r where the tangent line meets the curve again have respective parameters $-t/2$, $\pi - t/2$, and that the distance $pq = 4$.

3.3.3 Show that the speed $s(t)$ of the general trochoid (3.2) is given by the formula
$$s(t)^2 = (\lambda + 1)^2(h^2 - 2h\cos\lambda t + 1).$$

3.3.4 Show that the cycloid (3.3) has irregular parameters if and only if $h = 1$, and that in that case there are infinitely many.

3.3.5 Show that the cycloid (3.3) has self crossings if and only if $h > 1$, in which case there are infinitely many.

3.3.6 Determine those lines which are simultaneously tangent lines to the cycloid (3.3) at some parameter t_1, and normal to the cycloid at some parameter t_2.

4
Arc Length

The common thread in this chapter is the natural idea of arc length. Quite apart from being an interesting concept in its own right, there are specific applications relevant to the development of the subject. For instance in Section 4.2 we consider the question of changing the parameter on a regular curve, and in Section 4.3 prove that one can always choose a parameter with the property of 'unit speed'. Following the discussion of trochoids in the previous chapter, Section 4.4 provides a second excursion into the roulette concept, by looking at the idea of rolling a tangent line along a curve. That gives rise to the historically important concept of an 'involute' of a curve, which will turn out to be a reverse of the 'evolute' construction we will meet in Chapter 8.

4.1 Arc Length

The *speed* of the curve $z : I \to \mathbb{R}^2$ at the parameter t is defined to be the length of the tangent vector at t, i.e. the scalar defined by the formula

$$s(t) = |z'(t)| = \sqrt{x'(t)^2 + y'(t)^2}.$$

In complex number notation the tangent vector is $z'(t) = x'(t) + iy'(t)$ and the speed is simply its modulus $|z'(t)|$. Given scalars $a, b \in I$, we define the *arc length* from $t = a$ to $t = b$ to be the integral of the speed from a to b

$$l(a,b) = \int_a^b |z'(t)| dt.$$

Example 4.1 Let p, q be distinct points, and let L be the line joining them. Recall that L has a natural parametrization $z(t) = p + t(q-p)$ with p corresponding to $t = 0$, and q corresponding to $t = 1$. Then the length

4.1 Arc Length

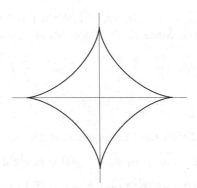

Figure 4.1. The astroid

of the line segment from $t = 0$ to $t = 1$ is, as we would hope,

$$\int_0^1 |z'(t)|dt = \int_0^1 |q - p|\, dt = |q - p| \int_0^1 dt = |q - p|.$$

Example 4.2 Let p, q be distinct points, and let L be the line joining them. We claim that *the line segment joining p, q is the shortest arc of any curve through p, q.* However obvious this may seem, it does require proof, providing a good application of the Cauchy Inequality. To this end consider *any* curve $z(t)$ with $p = z(a)$, $q = z(b)$. As a preliminary, note that for any fixed vector v we have

$$\int_a^b (v \bullet z'(t))dt = v \bullet \int_a^b z'(t)dt = v \bullet z(t)\Big|_a^b = v \bullet (q - p).$$

Now, using the Cauchy Inequality, and standard facts from calculus, we have

$$v \bullet (q - p) = \int_a^b (v \bullet z'(t))dt \le \int_a^b |v||z'(t)|dt = |v| \int_a^b |z'(t)|dt.$$

The claim follows from the result of the previous example, on taking v to be the unit vector defined by $v = (q - p)/|q - p|$.

Example 4.3 According to Exercise 3.3.1 the astroid is (up to a scalar multiple) the curve $x(t) = \cos^3 t$, $y(t) = \sin^3 t$. (Figure 4.1.) The speed is $|z'(t)| = \frac{3}{2}|\sin 2t|$, which vanishes if and only if t is an integer multiple of $\pi/2$: the traces of the irregular parameters are the four 'cusps' visible on the curve at $(1,0)$, $(0,1)$, $(-1,0)$, $(0,-1)$. We will calculate the arc length from $t = 0$ to $t = \pi/2$. Since $\sin 2t$ is non-negative on this interval, it is

$$\int_0^{\pi/2} |z'(t)|dt = \frac{3}{2}\int_0^{\pi/2} |\sin 2t|dt = \frac{3}{2}\int_0^{\pi/2} \sin 2t\, dt$$
$$= -\frac{3}{4}\Big[\cos 2t\Big]_0^{\pi/2} = \frac{3}{2}.$$

Example 4.4 Recall that the Cayley sextic is the curve with components

$$x(t) = \cos^3 t \cos 3t, \qquad y(t) = \cos^3 t \sin 3t$$

and has a self crossing at the point $p = (-1/8, 0)$ corresponding to the values $a = \pi/3$, $b = 2\pi/3$ of the parameter. (Example 2.11.) There are two loops starting and finishing at p, a smaller loop of length L, and a larger loop of length L'. (Figure 2.8.) The length L is easily determined. The curve has speed $3\cos^2 t$, so

$$L = \int_{\pi/3}^{2\pi/3} 3\cos^2 t\, dt = \frac{\pi}{2} - \frac{3\sqrt{3}}{4}.$$

Exercises

4.1.1   Let $z : I \to \mathbb{R}^2$ be a curve, and let a, b, c be scalars in I with $a < b < c$. Show that arc length is *additive*, in the sense that $l(a,c) = l(a,b) + l(b,c)$.

4.1.2   Show that the length of the catenary $x(t) = t$, $y(t) = \cosh t$ from $t = 0$ to $t = x$ is $\sinh x$.

4.1.3   Find the arc length of the astroid $x(t) = \cos^3 t$, $y(t) = \sin^3 t$ from $t = 0$ to $t = \pi$. (Compare with Example 4.3.)

4.1.4   Show that for $0 \leq x \leq \pi$ the length of the cardioid $z(t) = 2e^{it} - e^{2it}$ from $t = 0$ to $t = x$ is $8\sin(x/2)$.

4.1.5   Find an expression for the arc length of the cuspidal cycloid $x(t) = R(t - \sin t)$, $y(t) = R(1 - \cos t)$ from $t = 0$ to $t = t_0$ where $0 \leq t_0 \leq 2\pi$. Deduce that the arc length from $t = 0$ to $t = 2\pi$ is 8.

4.1.6   Show that the curve $x(t) = 3t^2$, $y(t) = t - 3t^3$ has a unique self crossing, determine the corresponding parameters a, b, and find the arc length from a to b.

4.1.7   Show that the arc length of the semicubical parabola $x(t) = 3at^2$, $y(t) = 2at^3$ with $a > 0$ measured from $t = 0$ is given by

$$s(t) = 2a\left\{(1+t^2)^{\frac{3}{2}} - 1\right\}.$$

4.1.8 Show that the arc length of the parabola $x(t) = at^2$, $y(t) = 2at$ with $a > 0$ measured from $t = 0$ is given by

$$s(t) = a\left\{t\sqrt{1+t^2} + \log\left(t + \sqrt{1+t^2}\right)\right\}.$$

4.1.9 Show that the arc length of the curve $x(t) = \cos^3 t$, $y(t) = \sin^3 t - 3\sin t$ from $t = 0$ to $t = 2\pi$ is 3π.

4.1.10 Show that the length of the arc of the curve $x(t) = \sinh t - t$, $y(t) = 3 - \cosh t$ cut off by the x-axis is

$$2\sqrt{2}\left\{2\sqrt{3} - \sqrt{2} - \log\frac{2+\sqrt{3}}{1+\sqrt{2}}\right\}.$$

4.1.11 Let λ be a positive real number. Find (in terms of λ) the two parameters a, b for which the curve defined by the following formulas meets the x-axis:

$$x(t) = t, \quad y(t) = \lambda e^t + \frac{1}{4\lambda e^t} - 2.$$

Show that the speeds of z at a, b are equal, and that the arc length from a to b is independent of λ.

4.1.12 The length L of the smaller loop for the Cayley sextic was determined in Example 4.4. Show that the length L' of the larger loop is given by

$$L' = \pi + \frac{3\sqrt{3}}{4}.$$

(Split the larger loop into two arcs, the first going from $t = 0$ to $t = \pi/3$, and the second from $t = 2\pi/3$ to $t = \pi$.)

4.2 Parametric Equivalence

We have already warned the reader to maintain a crystal clear distinction between the concept of a curve, and that of its trace. Two different curves may well have the same trace.

Example 4.5 The line $y = x$ can be parametrized as $x(t) = t$, $y(t) = t$: however, it is equally well parametrized as $x(t) = 2t$, $y(t) = 2t$ whose effect is to double the length of the tangent vector. There again, it could be parametrized as $x(t) = -t$, $y(t) = -t$ whose effect is to reverse the direction of the tangent vector, but leave its length unaltered.

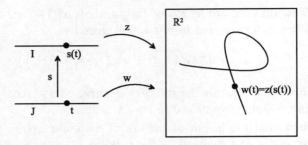

Figure 4.2. The idea of parametric equivalence

We need to be clear when two parametrizations give rise to the 'same' curve. Our starting point is to consider how to change the parameter on a curve. To this end we introduce the following definition. Let I, J be open intervals. A function $s : J \to I$ is a *change of parameter* when s is smooth, surjective, and has a non-zero derivative at every point. Recall from elementary calculus that a smooth function with an everywhere non-zero derivative is strictly monotone, and hence injective. Thus a change of parameters is necessarily a bijective map, so has an inverse $s^{-1} : I \to J$. Calculus tells us that the inverse is smooth, likewise with everywhere non-zero derivative, so is also a change of parameters.

Example 4.6 Let $I = J = \mathbb{R}$. The formula $s(t) = t + t^3$ defines a change of parameters. On the other hand the function $s(t) = t^3$ does not, since its derivative vanishes at $t = 0$.

Two curves $z : I \to \mathbb{R}^2$, $w : J \to \mathbb{R}^2$ are *parametrically equivalent* when there exists a change of parameter $s : J \to I$ with $w = z \circ s$, i.e. $w(t) = z(s(t))$ for all parameters t. One pictures the idea as in Figure 4.2. We say that w is a *reparametrization* of z. Note that parametrically equivalent curves have the *same* traces. It follows from the above remarks that parametric equivalence is an equivalence relation on curves.

Example 4.7 The curves $z(t) = (t, t^2)$, $w(t) = (2t, 4t^2)$ are parametrically equivalent under the change of parameters $s(t) = 2t$, since $z(s(t)) = z(2t) = (2t, 4t^2) = w(t)$.

Example 4.8 The 'positive' branch of the standard hyperbola has the parametrization $z(t) = (a \cosh t, b \sinh t)$ where a, b are positive real numbers. (Example 3.5.) The change of parameter $s(t) = e^t$ yields an

4.2 Parametric Equivalence

equivalent curve defined for $s > 0$

$$w(s) = \left(\frac{a}{2}\left(s + \frac{1}{s}\right), \frac{b}{2}\left(s - \frac{1}{s}\right)\right).$$

Example 4.9 Consider the rose curves $z(t) = 2be^{it}\cos nt$, where $b \neq 0$ and $n > 0$. (Example 2.7.) We claim that provided $n > 1$ the rose curves are parametrically equivalent to hypotrochoids. To see this, observe that under the change of parameter $s(t) = (n+1)t$ the formula becomes

$$z(s) = b\{e^{is} + e^{i(\frac{1-n}{1+n})s}\}.$$

That yields the general form (3.2) for a hypotrochoid when we set

$$R' = -b\left(\frac{n+1}{n-1}\right), \quad R = \frac{2bn}{n-1}, \quad h = \frac{n-1}{n+1}.$$

Many of the concepts we will meet in this book are invariant under parametric equivalence. For instance the concept of a 'regular parameter' is invariant in the following strict sense.

Example 4.10 Suppose $z : I \to \mathbb{R}^2$, $w : J \to \mathbb{R}^2$ are parametrically equivalent via the change of parameter $s : J \to I$. Thus $w(t) = z(s(t))$, for all parameters t. Differentiation using the Chain Rule gives $w'(t) = z'(s(t))s'(t)$, likewise for all t. Since s has a nowhere vanishing derivative, we deduce that $w'(t) \neq 0$ if and only if $z'(s(t)) \neq 0$. It follows that t is a regular parameter for w if and only if the corresponding value $s(t)$ is a regular parameter for z. Put another way, the regular parameters for w correspond one-to-one under s with the regular parameters for z. In particular, z is a regular curve if and only if w is regular.

Example 4.11 Numerous physical systems give rise to *simple harmonic motion* described by a function of the form $x(t) = a\sin(\omega t + \phi)$, where $a > 0$ is the *amplitude*, $\omega > 0$ is the *angular velocity*, and ϕ is the *phase constant*. Interesting situations arise when a particle in the plane is subject to two such motions, one in the x-direction, and the other in the y-direction. The particle then describes a *Lissajous figure* given parametrically as

$$x(t) = a_1\sin(\omega_1 t + \phi_1), \quad y(t) = a_2\sin(\omega_2 t + \phi_2). \tag{4.1}$$

For instance, the eight-curve of Example 2.9 is a Lissajous figure with $a_1 = a$, $a_2 = a/2$, $\omega_1 = 1$, $\omega_2 = 2$, $\phi_1 = \pi/2$, $\phi_2 = 0$. Such curves can be readily displayed on an oscilloscope screen, and exhibit a considerable diversity of behaviour, depending on the choices of constants. In view of

Figure 4.3. Some Lissajous figures

this complexity it is useful to simplify the parametric equations by making the change of parameters $s = \omega_2 t + \phi_2$ to get them into the following more convenient *standard form*, where $a = a_1$, $b = a_2$, $\omega = \omega_1/\omega_2$ and $\phi = \phi_1 - \omega\phi_2$:

$$x(s) = a\sin(\omega s + \phi), \quad y(s) = b\sin s.$$

Some examples of Lissajous figures are illustrated in Figure 4.3. One might ask when a Lissajous figure possesses an irregular parameter: in the special case when the phase constant $\phi = 0$, the required condition is that the angular velocity ω should be the quotient of two odd integers. (Exercise 4.2.3.)

Exercises

4.2.1 Show that the parametrizations $x(t) = \pm a \cosh t$, $y(t) = b \sinh t$ for the standard hyperbola (Example 3.5) are parametrically equivalent to $u(s) = \pm a \sec s$, $v(s) = b \tan s$, where s lies in the interval given by $-\pi/2 < s < \pi/2$.

4.2.2 By considering the change of parameter $s(t) = \log\left(\tan \frac{t}{2}\right)$ show that the tractrix of Example 2.22 is parametrically equivalent to the curve defined by the following formulas, where $0 < t < \pi$:
$$x(t) = \log\left(\tan \frac{t}{2}\right), \quad y(t) = \csc t.$$

4.2.3 Show that the Lissajous figure $x(s) = a \sin \omega s$, $y(s) = b \sin s$ has an irregular parameter if and only if ω is a quotient of two odd integers.

4.2.4 Let $a > 0$. Show that Agnesi's versiera (Example 3.7) is parametrically equivalent to the curve defined by the following formulas, where $-\pi/2 < t < \pi/2$:
$$x(t) = \frac{2a \cos t}{1 + \sin t}, \quad y(t) = a(1 + \sin t).$$

4.3 Unit Speed Curves

A *unit speed* curve is one for which the speed takes the constant unit value. For instance the standard parametrization $z(t) = e^{it}$ of the unit circle is a unit speed curve, since $|z'(t)| = |ie^{it}| = 1$. The point of the next result is that it will reduce a number of proofs to the case of unit speed curves, which are generally easier to handle. The key to the proof is the arc length concept.

Lemma 4.1 *Let $z : I \to \mathbb{R}^2$ be a regular curve, and let t_0 be a fixed choice of parameter. Then z is parametrically equivalent to a unit speed curve w, under a change of parameter $s : J \to I$ with $s(0) = t_0$ and everywhere positive derivative.*

Proof The arc length of z from t_0 to t is the smooth function $r : I \to \mathbb{R}$ defined by
$$r(t) = \int_{t_0}^{t} |z'(x)| dx.$$

Since z is regular we have $r'(t) = |z'(t)| > 0$ for all $t \in I$, so by calculus the image of r is an open interval J, and $r : I \to J$ is a change of parameter with $r(t_0) = 0$ having a positive derivative. The inverse function $s : J \to I$ is then a change of parameter with $s(0) = t_0$, having a positive derivative s' for which $s'(t)r'(s(t)) = 1$. The curve $w : J \to \mathbb{R}^2$ defined by $w(t) = z(s(t))$ is parametrically equivalent to z. Differentiating both sides of this relation gives $w'(t) = s'(t)z'(s(t))$, and taking moduli we obtain

$$|w'(t)| = |s'(t)||z'(s(t))| = |s'(t)||r'(s(t))| = |s'(t)r'(s(t))| = 1.$$

□

The value of this result is that for regular curves we know *in principle* that unit speed reparametrizations exist, though in practice it may be difficult to write them down explicitly. Here is a rather trivial example, where it is possible.

Example 4.12 Consider the parametrized circle $z(t) = z_0 + re^{it}$ with $r > 0$. We follow the proof of Lemma 4.1. Take $t_0 = 0$. Then $z'(t) = rie^{it}$, $|z'(t)| = r$, and a one line calculation gives $s(t) = rt$. Thus a unit speed reparametrization w of z is given by $w(s) = z_0 + re^{is/r}$.

Exercises

4.3.1 Find a unit speed reparametrization of the *equiangular spiral* $x(t) = e^t \cos t$, $y(t) = e^t \sin t$.

4.3.2 Let z, w be curves with the same domain I. Establish the identity

$$(z \bullet w)' = z \bullet w' + z' \bullet w.$$

What does this give in the case $z = w$? Use this result to show that for a unit speed curve z the tangent vector $z'(t)$ is orthogonal to $z''(t)$ at any parameter t.

4.4 Involutes

Here is another construction based on arc length. It represents our second excursion into the roulette concept. Recall the basic idea, that one curve (in a moving plane) is rolled on another (in a fixed plane) without slipping. Any point in the moving plane then traces a curve in the fixed plane. We discussed this construction in some detail in Section 3.3 when

4.4 Involutes

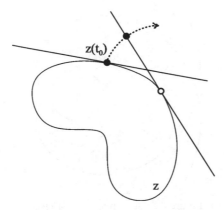

Figure 4.4. The involute construction

the fixed and the moving curves were circles, giving rise to the trochoids. Now consider the following situation. In the fixed plane we take any regular curve z. Choose some fixed *starting parameter* t_0. At t_0 there is a tangent line, which we think of as the curve in the moving plane. The tracing point is taken to be the point of contact $z(t_0)$ of z with its tangent line at t_0. We consider the locus of the marked tracing point as the tangent line rolls on z. Physically one thinks of the resulting curve as the path described by the end of a piece of string unwinding from z. (Figure 4.4.)

To obtain a formula for the locus of the tracing point we proceed as follows. As in Section 4.1, write $l(t_0, t)$ for the length of z from t_0 to t. Then at 'time' t we would expect the point $z(t_0)$ to become the point on the tangent line to z at t whose distance from $z(t)$ measured negatively from $z(t)$ is $l(t_0, t)$. Thus we are led to define the *involute* of z starting at t_0 to be the curve z^* defined by the following formula, where $T(t)$ is the unit tangent vector at t:

$$z^*(t) = z(t) - l(t_0, t) T(t). \tag{4.2}$$

Note that in the special case when z is of constant non-zero speed s we have $l(t_0, t) = s(t - t_0)$, and the formula for the involute simplifies to

$$z^*(t) = z(t) - (t - t_0) z'(t). \tag{4.3}$$

Example 4.13 The involute of the circle $x(t) = r \cos t$, $y(t) = r \sin t$ radius $r > 0$ centred at the origin is the curve known as the *Norwich*

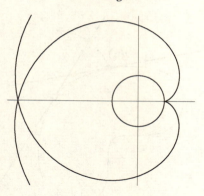

Figure 4.5. An involute of a circle

spiral. (Figure 4.5.) The circle parametrization has constant speed r, so we can apply (4.3). The reader will readily check that the involute z^* starting at $t_0 = 0$ has components

$$x^*(t) = r(\cos t + t \sin t), \quad y^*(t) = r(\sin t - t \cos t).$$

To maintain a clear mental picture, we have deliberately phrased the definition of the involute for *regular* curves. However, the formula makes sense for *any* curve z provided we restrict z^* to an interval on which z is regular. The point of that observation is that it allows us to extend the concept of involute to curves where the starting parameter t_0 is irregular.

Example 4.14 The cuspidal cycloid $z(t) = (t - \sin t, \cos t - 1)$ is obtained from that of Example 3.15 by setting $R = 1$ and reflecting in the x-axis. It is easily verified that the irregular parameters for z are those of the form $t = 2n\pi$ with n an integer, corresponding to the ends of the 'arches'. We will determine the involute of the arc $\pi < t < 2\pi$ starting at the regular parameter $t = \pi$. The tangent vector is

$$z'(t) = 2 \sin \frac{t}{2} \left(\sin \frac{t}{2}, -\cos \frac{t}{2} \right).$$

The speed and unit tangent vector are given by the following formulas:

$$|z'(t)| = 2 \left| \sin \frac{t}{2} \right|, \quad T(t) = \left(\sin \frac{t}{2}, -\cos \frac{t}{2} \right).$$

Furthermore, the length of the arc from π to t is

$$l(t) = \int_\pi^t |z'(x)| \, dx = \int_\pi^t 2 \sin \frac{x}{2} \, dx = 4 \left[-\cos \frac{x}{2} \right]_\pi^t = -4 \cos \frac{t}{2}$$

where, for the second equality, we use the fact that the integrand is non-negative on the interval $\pi < t < 2\pi$. It is now simply a matter of substituting in the formula (4.2) for z^* to see that it is given by

$$z^*(t) = (t + \sin t, -3 - \cos t) = z(t - \pi) + (\pi, -2).$$

Example 4.15 The crunch in the previous example is the last line of calculation, for it illustrates something quite remarkable. We obtain the involute of a cycloid by changing the parameter t to $t - \pi$, and then translating the resulting curve by the fixed vector $(\pi, -2)$. The physical interpretation is that the end point of a string of length 4 suspended from the point $(2\pi, 0)$ will describe another cycloid. That is the theoretical basis for Huyghens' *cycloidal pendulum*: it has the important physical property of being *isochronous*, meaning that the time taken for a single swing is independent of the amplitude.

Figure 4.5 suggests that the starting parameters for involutes of a circle are irregular. (And a minor calculation verifies that is the case.) However, that is a special case of a more general observation.

Example 4.16 Let z be a regular curve, and let z^* be the involute starting at the parameter t_0. For convenience, assume z is of unit speed, so the involute is $z^*(t) = z(t) - (t - t_0)z'(t)$. The derivative $-(t - t_0)z''(t)$ vanishes when $t = t_0$, so *the starting parameter t_0 is always irregular for the involute*. That can cause complications when trying to phrase general statements. For this reason it can be helpful to split the involute into two curves, namely the *forward involute* (the restriction of z^* to $t > t_0$) and the *backward involute* (the restriction of z^* to $t < t_0$).

Exercises

4.4.1 Determine the involutes of the semicubical parabola $x = 3t^2$, $y = 2t^3$. Show that for $X \leq 0$ there is exactly one involute through the point $(X, 0)$, and that when $X = -2$ the trace of this involute is a parabola.

4.4.2 Show that the involute of the catenary $x(t) = t$, $y(t) = \cosh t$ starting from $t = 0$ is the tractrix $x^*(t) = t - \tanh t$, $y^*(t) = \operatorname{sech} t$.

5
Curvature

In this chapter we introduce one of the fundamental ideas of differential geometry, namely the 'curvature' of a curve. In Chapter 6 we will see that there is a formal sense in which the speed and curvature completely determine the curve, up to a natural equivalence relation. The concept will enable us to study various special points on curves which play an important role in understanding the geometry of the curve.

5.1 The Moving Frame

Suppose t is a *regular* parameter of a curve z, so both the tangent and the normal vectors are non-zero. By the *unit tangent vector* at t we mean the unit vector $T(t)$ in the same direction as the tangent vector, and by the *unit normal vector* at t we mean the unit vector $N(t)$ in the same direction as the normal vector. Thus, writing $s(t)$ for the speed, we have

$$T(t) = \left(\frac{x'(t)}{s(t)}, \frac{y'(t)}{s(t)}\right), \quad N(t) = \left(-\frac{y'(t)}{s(t)}, \frac{x'(t)}{s(t)}\right). \tag{5.1}$$

The unit vectors $T(t)$, $N(t)$ form an orthonormal basis for the plane, called the *frame* at t. One of the basic pictures of the subject is that as t varies so this frame changes with t, giving rise to the idea of a *moving frame*. The intuitive picture is that at each instant t we translate the origin to the point $z(t)$: then as t varies we can picture the frame moving along the curve, as illustrated in Figure 5.1 for an arc of a circle.

The idea of a frame is independent of the parametrization, in the sense of the following lemma: the proof is left to the reader.

Lemma 5.1 *Let $z_1 : I_1 \to \mathbb{R}^2$, $z_2 : I_2 \to \mathbb{R}^2$ be parametrically equivalent curves via a change of parameter $s : I_2 \to I_1$. Write T_1, N_1 (respectively T_2,*

5.1 The Moving Frame

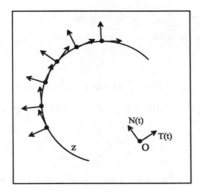

Figure 5.1. The idea of a moving frame

N_2) for the unit tangent and normal vectors to z_1 (respectively z_2). Then $T_2(t) = T_1(s(t))$, and hence $N_2(t) = N_1(s(t))$. Thus the frame for z_1 at any regular parameter t coincides with the frame for z_2 at the corresponding parameter $s(t)$.

We wish now to determine the 'rate of change' of our moving frame $T(t)$, $N(t)$ as t varies, i.e. the derivatives $T'(t)$, $N'(t)$. Since $T(t)$, $N(t)$ form a basis for $\mathbb{R}^2$, we can write

$$T'(t) = aT(t) + bN(t), \quad N'(t) = cT(t) + dN(t) \qquad (5.2)$$

for unique scalars a, b, c, d, depending on t. To determine these coefficients, take scalar products of both sides of these relations with $T(t)$, $N(t)$ to obtain

$$a = T(t) \bullet T'(t), \ b = N(t) \bullet T'(t), \ c = T(t) \bullet N'(t), \ d = N(t) \bullet N'(t).$$

We can evaluate these expressions as follows. The fact that $T(t)$, $N(t)$ are both of unit length and orthogonal is expressed by the relations

$$T(t) \bullet T(t) = 1, \quad N(t) \bullet N(t) = 1, \quad T(t) \bullet N(t) = 0.$$

Differentiating with respect to t (using the formula for differentiating a product) we deduce immediately that $a = 0$, $d = 0$, $b + c = 0$. Setting $b = \omega(t)$ we see that (5.2) reduces to the *Serret–Frenet* Formulas

$$T'(t) = \omega(t)N(t), \quad N'(t) = -\omega(t)T(t).$$

Bear in mind that these formulas only make sense for regular values of the parameter t.

Exercises

5.1.1 Write out a detailed proof of Lemma 5.1.

5.2 The Concept of Curvature

The *curvature* $\kappa = \kappa(t)$ of the curve z at a regular parameter t is defined to be

$$\kappa(t) = \frac{\omega(t)}{s(t)} = \frac{T'(t) \bullet N(t)}{s(t)}.$$

Note that with this definition we can re-write the Serret–Frenet Formulas in the following form:

$$T'(t) = \kappa(t)s(t)N(t), \quad N'(t) = -\kappa(t)s(t)T(t). \tag{5.3}$$

In the case of a unit speed curve these formulas simplify to

$$T'(t) = \kappa(t)N(t), \quad N'(t) = -\kappa(t)T(t). \tag{5.4}$$

The import of the next result is that the concept of curvature is preserved by parametric equivalence, in the following precise sense.

Lemma 5.2 *Let z_1, z_2 be regular curves with domains I_1, I_2 and curvature functions κ_1, κ_2. Assume the curves are parametrically equivalent via a change of parameter $s : I_2 \to I_1$ having positive derivative. Then $\kappa_2(t) = \kappa_1(s(t))$.*

Proof Write T_1, N_1, s_1 (respectively T_2, N_2, s_2) for the unit tangent vector, unit normal vector and speed of z_1 (respectively z_2). The curves z_1, z_2 are related by $z_2(t) = z_1(s(t))$. Differentiating this relation, and taking moduli of both sides, we see that $s_2(t) = s'(t)s_1(s(t))$. (That uses the hypothesis that s has a *positive* derivative.) By Lemma 5.1 we have $T_2(t) = T_1(s(t))$, $N_2(t) = N_1(s(t))$ and differentiation of the former relation yields $T_2'(t) = s'(t)T_1'(s(t))$. We now have

$$\begin{aligned}\kappa_2(t) &= \frac{T_2'(t) \bullet N_2(t)}{s_2(t)} \\ &= \frac{s'(t)T_1'(s(t)) \bullet N_1(s(t))}{s'(t)s_1(s(t))} \\ &= \frac{T_1'(s(t)) \bullet N_1(s(t))}{s_1(s(t))} = \kappa_1(s(t)).\end{aligned}$$

$\square$

5.3 Calculating the Curvature

Combining this result with Lemma 4.1 we see that in discussing the curvature of regular curves we can tacitly assume our curves are of unit speed. Here is an application of this fact, providing a good mental picture of the curvature concept.

Lemma 5.3 *Let z be a unit speed curve, let u be any fixed unit vector in the plane, and let t_0 be a fixed value of the parameter. There exists a smooth function $\theta(t)$, defined for t sufficiently close to t_0, representing the angle between u and the tangent vector $T(t)$. Moreover the function $\theta(t)$ satisfies the relation $\kappa(t) = \theta'(t)$: paraphrased, the curvature is the rate of change of the angle between $T(t)$ and the fixed direction u.*

Proof The existence of the function $\theta(t)$ with the desired properties follows from the fact that, for t sufficiently close to a given value t_0, at least one of the equations

$$u \bullet T(t) = \cos\theta(t), \quad -u \bullet N(t) = \sin\theta(t)$$

has a smooth solution $\theta(t)$. (Recall that the inverses of the sine and cosine functions are smooth: of course $\theta(t)$ is only defined up to multiples of 2π.) Differentiating these relations with respect to t, and using the Serret–Frenet Formulas (5.4), we obtain

$$-\theta'(t)\sin\theta(t) = u \bullet T'(t) = \kappa(t)u \bullet N(t) = -\kappa(t)\sin\theta(t)$$

$$\theta'(t)\cos\theta(t) = -u \bullet N'(t) = \kappa(t)u \bullet T(t) = \kappa(t)\cos\theta(t).$$

Since one of $\sin\theta(t)$, $\cos\theta(t)$ is non-zero we deduce that $\kappa(t) = \theta'(t)$, as was claimed. $\square$

5.3 Calculating the Curvature

In practice it is convenient to have a formula for the curvature $\kappa(t)$ in terms of the components $x(t)$, $y(t)$ of a curve $z(t)$ and their derivatives. Differentiating the formula (5.1) for the unit tangent vector T we obtain

$$T' = \left(\frac{sx'' - x's'}{s^2}, \frac{sy'' - y's'}{s^2}\right)$$

where s denotes the speed. Taking the scalar product of this expression with the unit normal vector N produces

$$T' \bullet N = \frac{x'y'' - x''y'}{s^2}.$$

We obtain the required expression for the curvature κ on dividing by s, yielding the explicit formula

$$\kappa = \frac{x'y'' - x''y'}{(x'^2 + y'^2)^{\frac{3}{2}}}. \tag{5.5}$$

Example 5.1 Consider a general line segment $x(t) = \alpha + \beta t$, $y(t) = \gamma + \delta t$ with at least one of β, δ non-zero, where t lies in some open interval I. Here $x'(t) = \beta$, $x''(t) = 0$, $y'(t) = \delta$, $y''(t) = 0$ and on substituting in (5.5) we obtain $\kappa(t) = 0$. Thus line segments have constant zero curvature.

Example 5.2 Consider an arc of a general circle $x(t) = a + r\cos t$, $y(t) = b + r\sin t$ with centre (a, b) and radius $r > 0$, where t lies in some open interval I. Here $x'(t) = -r\sin t$, $x''(t) = -r\cos t$, $y'(t) = r\cos t$, $y''(t) = -r\sin t$, and substitution in (5.5) gives $\kappa(t) = 1/r$. Thus arcs of circles are curves of constant non-zero curvature.

Example 5.3 Consider the ellipse $x(t) = a\cos t$, $y(t) = b\sin t$, with $0 < a < b$. Here we have $x'(t) = -a\sin t$, $x''(t) = -a\cos t$, $y'(t) = b\cos t$, $y''(t) = -b\sin t$, and substitution in (5.5) yields the formula below. Note that in the limiting case $a = b = r$, when the ellipse becomes a circle of radius r, the formula agrees (as it should) with that given in Example 5.2 for the curvature of a circle.

$$\kappa(t) = \frac{ab}{(a^2 \sin^2 t + b^2 \cos^2 t)^{\frac{3}{2}}}.$$

Example 5.4 Consider the graph $y = f(x)$ of a smooth function $f(x)$ viewed as the curve $x(t) = t$, $y(t) = f(t)$. In that case $x'(t) = 1$, $x''(t) = 0$, $y'(t) = f'(t)$, $y''(t) = f''(t)$, and substitution in (5.5) gives

$$\kappa(t) = \frac{f''(t)}{(1 + f'(t)^2)^{\frac{3}{2}}}.$$

Here are two applications of the Serret–Frenet Formulas. We saw above that line segments and arcs of circles are examples of curves of constant curvature. The next two results show that these are the only possible curves of constant curvature. In fact these results follow from a Uniqueness Lemma we will prove in Chapter 6. However, we insert

5.3 Calculating the Curvature

them here because they illustrate well the way in which the Serret–Frenet Formulas can be applied.

Lemma 5.4 *Any regular parametrization z of a line segment has constant zero curvature κ. Conversely, any regular curve z of constant zero curvature κ is a line segment.*

Proof Suppose first that z has trace a line segment, so $z \bullet n = -c$ for some unit vector n, and some scalar c. (Example 1.8.) Differentiation yields the identity $z' \bullet n = 0$. That implies T is constant, and hence $T' = 0$ identically. The Serret–Frenet Formula $T' = \kappa s N$, where s is the speed, then shows that $\kappa = 0$. Conversely, suppose z has constant zero curvature κ. The Serret–Frenet Formula $N' = -\kappa s T$ then gives $N' = 0$, so N is a constant unit vector, and $(z \bullet N)' = z \bullet N' + z' \bullet N = sT \bullet N = 0$: thus $z \bullet N = -c$, for some constant c, which is the equation of a line. (Example 1.8.) □

Lemma 5.5 *Any regular parametrization z of an arc of a circle of radius $r > 0$ has curvature of constant absolute value $|\kappa| = 1/r$. Conversely, the trace of any regular curve z of constant non-zero curvature κ is an arc of a circle of radius r given by $|\kappa| = 1/r$.*

Proof Suppose first that z has trace an arc of a circle, so there exist a vector p, and a real number $r > 0$ such that $|z(t) - p| = r$ for all t. Squaring both sides and differentiating we get $(z(t)-p) \bullet z'(t) = 0$ for all t. Thus $N(t)$ is a multiple of $z(t) - p$, indeed $\pm N(t) = (z(t)-p)/r$, and hence $\pm N'(t) = z'(t)/r$. On the other hand $T(t) = z'(t)/s(t)$ where $s(t)$ denotes the speed. Substituting in the Serret–Frenet Formula $N' = -\kappa s T$ we see that $\kappa(t) = \pm 1/r$, and hence $|\kappa| = 1/r$, as required. Conversely, suppose that the regular curve z has constant non-zero curvature κ. Using the Serret–Frenet Formula $N' = -\kappa s T$ we obtain

$$\left(z + \frac{N}{\kappa}\right)' = z' + \frac{N'}{\kappa} = sT + \left(\frac{-\kappa s T}{\kappa}\right) = 0.$$

Integrating, we see that for some constant vector p we have $z + N/\kappa = p$ and hence $|z - p| = 1/|\kappa|$, which is the equation of a circle of radius $1/|\kappa|$ centre p. □

Exercises

5.3.1 In each of the following cases find the curvature of the given curve.

(i) $x(t) = t,$ $y(t) = t^3$
(ii) $x(t) = t,$ $y(t) = t^4$
(iii) $x(t) = t^2,$ $y(t) = t^3$
(iv) $x(t) = t^2 - 1,$ $y(t) = t^3 - t$
(v) $x(t) = t^2 + 1,$ $y(t) = t^3 + t$

5.3.2 Find a formula for the curvature of the parabola $x = at^2$, $y = 2at$ with $a > 0$. Show that the vertex is the unique point on the parabola where the curvature assumes a maximal value. Show that for any positive real number k with $k < 1/2a$ there are exactly two distinct points on the parabola for which the curvature assumes the value k.

5.3.3 Let $P = (a, b)$ where $a^2 > b$, and let A, B be the two points on the parabola $x(t) = t$, $y(t) = t^2$ where the tangent passes through P. Prove that the ratio $PA^3 : PB^3 = \kappa_B : \kappa_A$ where κ_A, κ_B are the curvatures of the parabola at A, B respectively.

5.3.4 In Example 5.3 it was shown that the ellipse $x(t) = a\cos t$, $y(t) = b\sin t$, with $0 < a < b$, has curvature $\kappa = ab/\phi^{3/2}$, where ϕ is the function defined by

$$\phi(t) = a^2 \sin^2 t + b^2 \cos^2 t.$$

Show that the maximum and minimum values of the curvature are assumed at the vertices. What are their values? Show that ϕ is strictly increasing for $0 \le t \le \pi/2$. Deduce that for any real number k with $b/a^2 < k < a/b^2$ there are exactly four distinct points on the ellipse with $\kappa = k$.

5.3.5 Let p be a point on the ellipse $x(t) = a\cos t$, $y(t) = b\sin t$ and let q be a point on the x-axis lying on the normal line at p. Write d for the distance between p, q. Show that the radius of curvature ρ at p is given by the formula $\rho = cd^3$ where $c = a^2/b^4$.

5.3.6 Find a formula for the curvature of the limacon defined by the formula $z(t) = 2e^{it} + he^{2it}$.

5.3.7 Show that the curvature of the cuspidal cycloid $x(t) = t - \sin t$, $y(t) = 1 - \cos t$ at a regular value t is given by

$$\kappa(t) = \frac{-1}{4\sin(t/2)}.$$

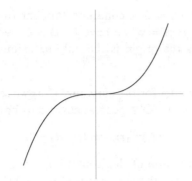

Figure 5.2. A typical inflexion

5.3.8 Let z be a unit speed curve, and let t be a parameter for which $z'(t) = 1$. Show that the curvature $\kappa(t) = \Im z''(t)$.

5.3.9 Let $f(x)$ be a smooth function with $f(0) = 0$, $f'(0) = 0$. Show that the curvature of the graph $y = f(x)$ is given by the *Newton formula*

$$\kappa(0) = \lim_{x \to 0} \frac{2y}{x^2}.$$

5.4 Inflexions

In Lemma 5.3 we saw that for a unit speed curve z the curvature $\kappa(t)$ represented the rate of change of the angle $\varphi(t)$ between the tangent vector $z'(t)$ and any given direction. In particular, at points where the curve is infinitesimally 'flat' we expect the tangent direction to be (instantaneously) constant, and therefore the curvature to vanish. That motivates the following definition. A regular parameter t for z is *inflexional* when $\kappa(t) = 0$, and then the corresponding point $z(t)$ on the trace is an *inflexion*. The mental picture is that typically the curve will bend one way on one side of the inflexion, and the other way on the other side: Figure 5.2 illustrates this typical behaviour for the curve $x(t) = t$, $y(t) = t^3$ with $\kappa(0) = 0$. In Chapter 7 we will see that there are other ways of viewing this basic geometric idea.

Example 5.5 Consider the graph $y = f(x)$ of a smooth function $f(x)$ viewed as the curve $x(t) = t$, $y(t) = f(t)$. In view of the formula for the curvature given in Example 5.4 inflexions appear precisely when the

70   *Curvature*

second derivative $f''(t) = 0$, a condition familiar from school geometry. For instance, when $f(x) = x^3$ we have $f''(x) = 6x$, which vanishes if and only if $x = 0$. Thus the origin is the only inflexion of the curve $x = t$, $y = t^3$.

In view of the formula for the curvature (5.5) we see that the condition for a *regular* parameter t of a general curve z to be inflexional is that

$$x'(t)y''(t) - x''(t)y'(t) = 0. \tag{5.6}$$

An alternative formulation of this condition, useful when using complex number notation, is that the derivatives $z'(t)$, $z''(t)$ should be linearly dependent. Here are some examples, illustrating the mechanics of analysing this condition.

Example 5.6 Agnesi's versiera was introduced in Example 3.7 as a regular curve of the form

$$x(t) = 2at, \quad y(t) = \frac{2at}{1 + t^2}$$

where a is a positive constant. The reader is invited to check that the left hand side of (5.6) vanishes if and only if $3t^2 = 1$, producing two inflexions lying symmetrically about the y-axis. (Figure 3.4.)

Example 5.7 The piriform was introduced in Example 2.16 as a parametrized curve of the form

$$x(t) = a(1 + \cos t), \quad y(t) = b \sin t \, (1 + \cos t)$$

where a, b are positive constants. We verified that irregular parameters were defined by $\sin t = 0$, $\cos t = -1$ giving the 'cusp' point on the trace. (Figure 2.12.) The reader will readily check that

$$x'y'' - x''y' = -ab(C + 1)(2C^2 - 2C - 1)$$

with $C = \cos t$ necessarily in the range $-1 \le C \le 1$. The displayed expression vanishes when one of the following relations is satisfied:

$$C = -1, \quad C = \frac{1 + \sqrt{3}}{2}, \quad C = \frac{1 - \sqrt{3}}{2}.$$

The first gives the 'cusp' point, the second is outside the allowable range, and the third gives two solutions, in the range $\pi < t < 2\pi$. The piriform has therefore two inflexions, as Figure 2.12 suggests.

5.4 Inflexions

Example 5.8 Consider the epicycloids and hypocycloids defined by the formula

$$z(t) = (\lambda + 1)e^{it} - e^{i(\lambda+1)t}.$$

Assuming $\lambda \neq -1$, it was shown in Example 3.10 that a parameter t is irregular if and only if $e^{i\lambda t} = 1$. A regular parameter t is inflexional if and only if the derivatives

$$z'(t) = (\lambda + 1)ie^{it}w, \quad z''(t) = (\lambda + 1)e^{it}\{\lambda - (\lambda + 1)w\}$$

are linearly dependent, i.e. if and only if iw, $\lambda - (\lambda + 1)w$ are linearly dependent. Direct calculation shows that this condition is satisfied if and only if $\lambda = -2$, which is the case of Cardan circles. (Example 3.12.) Thus there are no inflexional parameters, except in the case of Cardan circles, when *every* parameter is inflexional, and the trace is a diameter of the fixed circle.

Example 5.9 We will find the inflexions of the family of limacons $z(t) = 2e^{it} - he^{2it}$, where h is a positive real number. (Example 3.13.) By computation

$$x'(t)y''(t) - x''(t)y'(t) = 4\{2h^2 - 3h\cos t + 1\} \tag{5.7}$$

In Example 3.10 it was shown that the curve is regular if and only if $h \neq 1$. In the exceptional case $h = 1$ of a cardioid there is just one irregular point, determined by $\cos t = 1$. When $h = 1$ the above expression vanishes only at the irregular point, and there are no inflexions. Suppose now that $h \neq 1$. We obtain inflexions if and only if the expression (5.7) vanishes, i.e. if and only if there exists a parameter t for which

$$\cos t = H = \frac{1 + 2h^2}{3h}. \tag{5.8}$$

Such a relation places a restriction on h, since it holds if and only if $-1 \leq H \leq 1$. But $H > 0$ since h is assumed to be positive, so the first inequality is automatic. The second inequality is equivalent to $0 \geq 2h^2 - 3h + 1 = (2h - 1)(h - 1)$, and holds if and only if $1/2 \leq h \leq 1$, with $H = 1$ if and only if $h = 1/2$ or $h = 1$. Since we are supposing $h \neq 1$, we deduce that there are inflexions if and only h lies in the range $1/2 \leq h < 1$. When $h = 1/2$ the parameter t is inflexional if and only if $\cos t = 1$, giving just one solution in the range $0 \leq t < 2\pi$, namely $t = 0$: thus we obtain just one inflexion. When $1/2 < h < 1$ we have $0 < H < 1$ and a glance at the graph of $\cos t$ will convince the reader that

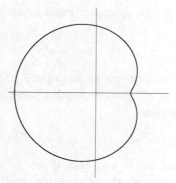

Figure 5.3. A biflexional limacon

in the range $0 \le t < 2\pi$ there are exactly two solutions t_1, t_2 for (5.8): thus we obtain exactly two inflexions. For this reason the limacons with $1/2 < h < 1$ are said to be *biflexional*. (Figure 5.3.)

Exercises

5.4.1 Show that none of the curves parametrizing the standard conics have inflexional parameters.

5.4.2 Show that the curve defined by $x(t) = 3t^2 - 1$, $x(t) = 3t^3 - t$ has no inflexional parameters.

5.4.3 Calculate the curvature of the curve $x(t) = t - \sinh t \cosh t$, $y = 2\cosh t$, and deduce that the curve has no inflexions.

5.4.4 In each of the following cases show that the curve has just one irregular parameter, and just one inflexional parameter.

(i) $x(t) = t^2$, $y(t) = t^5 - t^4$
(ii) $x(t) = t^2 - t^3$, $y(t) = t^5$
(iii) $x(t) = t^2 + t^4$, $y(t) = t^2 + t^5$.

5.4.5 Let a be a positive constant. Show that the eight-curve given by $x(t) = a\cos t$, $y(t) = a\cos t \sin t$ is regular. Find a formula for the curvature function. Show that the curve has two inflexional parameters in the range $0 < t < 2\pi$, giving rise to inflexions at the self crossing.

5.4.6 Show that none of the four 'branches' of the cross curve have inflexional parameters. (Example 2.21.)

5.5 Limiting Behaviour

5.4.7 The *Kampyle of Eudoxus* is the curve defined by $x(t) = a\sec t$, $y(t) = a\sec t \tan t$ where $a > 0$ and $-\pi/2 < t < \pi/2$. (It arises in connexion with the problem of duplicating the cube.) Show that the Kampyle has exactly two inflexions.

5.4.8 Let $r > 0$ and let $\gamma = \alpha + i\beta$ where α, β are real numbers with $\beta \neq 0$. Show that the curvature of the standard equiangular spiral $z(t) = re^{\gamma t}$ is given by the following formula. Deduce that z has no inflexions.

$$\kappa(t) = \left\{ \frac{\beta}{r|\gamma|} \right\} e^{-\alpha t}.$$

5.4.9 A curve z has the property that its components x, y satisfy the relation $y(t) = tx(t)$. Show that a regular parameter t is inflexional if and only if $2x'(t)^2 = x(t)x''(t)$. Use this relation to show that in each of the following cases the curve has no inflexional parameters.

(i) $x(t) = \dfrac{2at^2}{1+t^2}$, (ii) $x(t) = \dfrac{1-t^2}{1+t^2}$, (iii) $x(t) = \dfrac{t^2-1}{3t^2+1}$.

Incidentally (i) is the cissoid of Diocles, (ii) is the right strophoid, and (iii) is Descartes' folium.

5.5 Limiting Behaviour

Although the curvature $\kappa(t)$ of a curve z is not defined at an irregular parameter t_0, it is quite possible that it may tend to a limit as $t \to t_0$ through regular parameters t. However in that case a detailed analysis is required to determine the value of the limit.

Example 5.10 The three curves z_1, z_2, z_3 below all have exactly one irregular parameter, namely $t = 0$. Figure 5.4 illustrates the nature of the curves near the origin.

$$z_1(t) = (t^2, t^3), \quad z_2(t) = (t^2, t^5), \quad z_3(t) = (t^2 + t^3, t^4).$$

The reader is invited to verify that the respective curvature functions κ_1, κ_2, κ_3 have the following forms:

$$\kappa_1(t) = \frac{3}{4t} + O(2), \quad \kappa_2(t) = \frac{15t}{4} + O(2), \quad \kappa_3(t) = 2 + O(1)$$

where $O(n)$ denotes terms in t of degree $\geq n$. As $t \to 0$ so $\kappa_1 \to \infty$, $\kappa_2 \to 0$ and $\kappa_3 \to 2$. Note however that the illustrations of the curves

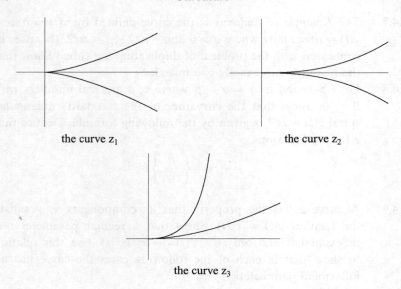

Figure 5.4. Forms of the curves near the origin

near the origin look remarkably similar: the curvature function is picking up information which the eye does not.

Exercises

5.5.1 Investigate the limiting value of the curvature as $t \to 0$ for the curve $x(t) = t^2$, $y(t) = t^4 + t^5$.

5.5.2 Investigate the limiting value of the curvature as $t \to 0$ for the curve $x(t) = t^2 + t^4$, $y(t) = t^2 + t^5$.

6
Existence and Uniqueness

We know that a curve determines a curvature function, providing valuable geometric information about the original curve. A natural question to ask is whether there exist curves with prescribed curvature, i.e. given a smooth function κ is there a regular curve z whose curvature function is κ? Indeed, this is the case.

Theorem 6.1 *Let $\kappa : I \to \mathbb{R}$ be a smooth function. There exists a regular parametrized curve $z : I \to \mathbb{R}^2$ whose associated curvature function is κ.* (The Existence Theorem.)

Proof Let $\theta : I \to \mathbb{R}$ be any function with the property that $\theta'(t) = \kappa(t)$ for all $t \in I$. Further, let $x(t)$, $y(t)$ be smooth functions with $x'(t) = \cos \theta(t)$, $y'(t) = \sin \theta(t)$, and let $z : I \to \mathbb{R}^2$ be the curve with those components. Clearly z is a unit speed curve. The associated tangent and normal vectors are

$$T(t) = (\cos \theta(t), \sin \theta(t)), \quad N(t) = (-\sin \theta(t), \cos \theta(t)).$$

Differentiating T we obtain the following relation, from which it is immediate that κ is the curvature function for z:

$$T'(t) = \theta'(t)(-\sin \theta(t), \cos \theta(t)) = \kappa(t)N(t).$$

$\square$

However, that raises the question of the extent to which curves with a given curvature function are 'unique'. In Section 6.4 we introduce the notion of 'congruence' for curves, and establish that congruent curves necessarily have the same curvature function. The main result of this chapter is the Uniqueness Theorem, establishing a converse, namely that

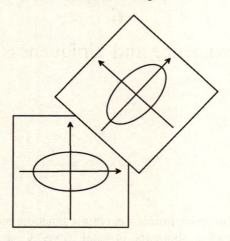

Figure 6.1. Superimposing two curves

if two regular curves have the same speeds and curvatures respectively then they are congruent. In other words, up to the relation of congruence, a curve is completely determined by its speed and curvature. This result allows us to recover facts established by direct calculation, for instance the fact that the only curves with constant curvature functions are line segments and arcs of circles.

6.1 Isometries

Our starting point is the need to be clear about what we mean by two curves being 'the same'. The idea is that it should be possible to superimpose the one trace on the other. Let us clarify this. Start from a picture in which the two curves are traced on the same plane, with the axes drawn in. Now, mentally separate the curves, by imagining the two curves traced on two separate planes (with the two sets of axes superimposed): one fixed plane, and on top of it a moving plane. The planes could be realized as plastic transparencies, one resting on the other. Each transparency has axes drawn in, and a curve traced on it. (Figure 6.1.) Then the idea is that the curves are 'the same' when we can place the moving plane on the fixed plane in such a way that the two curves are perfectly superimposed.

To make this intuition precise, think of the moving plane as the image of the fixed plane under a mapping which 'preserves' the Euclidean

6.1 Isometries

structure, in the following precise sense. By a *planar* mapping we mean a mapping $I : \mathbb{R}^2 \to \mathbb{R}^2$: an *isometry* is a surjective planar mapping I which *preserves distance*, in the sense that for all vectors u, v

$$|I(u) - I(v)| = |u - v|.$$

The surjectivity condition is actually redundant: it can be deduced from the distance preserving condition. The reason for inserting the condition is that 'isometries' can be defined in wider contexts within mathematics (metric spaces) where the surjectivity condition is no longer redundant, so it is sensible to have a uniform definition. Note that *isometries are automatically injective*: indeed if $I(u) = I(v)$ then $0 = |I(u) - I(v)| = |u - v|$ so $u = v$. Thus isometries are bijective mappings of the plane, and possess inverses.

Example 6.1 By a *translation* T we mean a planar mapping T given (in complex notation) by a formula $T(z) = z + b$ where $b = (u, v)$ is a fixed vector. Such a mapping is surjective: indeed any point w can be written in the form $w = T(z)$ by choosing $z = w - b$. We think of a translation as a sliding of the plane over itself through b. In terms of components, T is the map $(x, y) \to (X, Y)$ where $X = x + u$, $Y = y + v$. Translations are isometries. Indeed for any two points z_1, z_2 we have

$$|T(z_1) - T(z_2)| = |(z_1 + b) - (z_2 + b)| = |z_1 - z_2|.$$

Example 6.2 In exactly the same way, the reader will verify that any planar mapping I given by a formula $I(z) = uz + b$, with u unit, is an isometry: likewise, any planar mapping I of the form $I(z) = u\bar{z} + b$, with u unit, is an isometry. (Exercise 6.1.1.) In fact we will see that *any* isometry has one of these two forms.

It is profitable to phrase the basic properties of isometries in the language of elementary group theory. Recall first that the bijective mappings of *any* set form a group under the operation of composition. In particular the bijective mappings of the plane form a group, and the isometries form a subgroup, known as the *Euclidean group* $E(2)$. That means that the identity mapping is an isometry, that the composite of two isometries is an isometry, and that the inverse of an isometry is another isometry. (Exercise 6.1.2.) Here is a very simple lemma elucidating the structure of general isometries.

Lemma 6.2 *Any isometry I can be written as the composite $I = T \circ R$ of a translation T and an isometry R with the property that $R(0) = 0$. Moreover, T and R are unique.*

Proof Write $T(z) = z + I(0)$, $R(z) = I(z) - I(0)$. Then T is a translation, R is an isometry, and $I = T \circ R$. For uniqueness, suppose that $I = T_1 \circ R_1$, $I = T_2 \circ R_2$, where R_1, R_2 are isometries with $R_1(0) = 0$, $R_2(0) = 0$, and T_1, T_2 are translations given by formulas $T_1(z) = z + b_1$, $T_2(z) = z + b_2$. Then for all z we have $I(z) = R_1(z) + b_1 = R_2(z) + b_2$. Setting $z = 0$ gives $b_1 = b_2$, hence $T_1 = T_2$: but then $R_1(z) = R_2(z)$ for all z, giving $R_1 = R_2$. □

Isometries R with $R(0) = 0$ have properties not enjoyed by general isometries. For instance, they preserve not just distance, but length and scalar product.

Lemma 6.3 *Any isometry R with $R(0) = 0$ preserves length, in the sense that $|R(u)| = |u|$ for all vectors u. Moreover R preserves scalar products, in the sense that $R(u) \bullet R(v) = u \bullet v$ for all vectors u, v.*

Proof For the first statement we have simply to note that since $R(0) = 0$ we have

$$|R(u)| = |R(u) - 0| = |R(u) - R(0)| = |u - 0| = |u|.$$

For the second, recall the Polarization Identity of Exercise 1.3.2. Applying the identity firstly to u, v, and secondly to $R(u), R(v)$, gives

$$2(u \bullet v) = |u|^2 + |v|^2 - |u - v|^2$$
$$2(R(u) \bullet R(v)) = |R(u)|^2 + |R(v)|^2 - |R(u) - R(v)|^2.$$

The result now follows from the fact that R preserves both length and distance. □

Exercises

6.1.1 Let u, b be complex numbers with u unit. Show that the planar mapping I given by a formula $I(z) = uz + b$ is an isometry: likewise, show that the planar mapping I given by the formula $I(z) = u\bar{z} + b$ is an isometry.

6.1.2 Show that the isometries form a group under the operation of composition, and that the translations form a subgroup.

6.2 Fixed Points and Formulas

6.1.3 Let I be an isometry. Show that the image under I of a line is another line. Also, show that the image under I of a circle of radius r and centre c is a circle of radius r and centre $I(c)$.

6.1.4 A *similarity* is a surjective planar mapping S for which there exists a positive real number k (the *scaling factor*) such that $|S(u) - S(v)| = k|u - v|$ for all choices of vectors u, v. Show that similarities are bijective mappings of the plane. Show that the similarities form a group under the operation of composition.

6.1.5 A *central dilation* with *centre* w and *scaling factor* k is a planar mapping D given by a formula $D(z) = w + k(z - w)$ where w is a fixed point, and $k > 0$. Verify that a central dilation is a similarity with scaling factor k. Show that any similarity S (with scaling factor k) can be written as the composite of an isometry I and a central dilation D with centre the origin (and scaling factor k).

6.2 Fixed Points and Formulas

By a *fixed point* of an isometry I we mean a point p for which $I(p) = p$. This key idea leads us very quickly to an understanding of the structure of isometries. Here is the basic result.

Lemma 6.4 *For any isometry I the set of fixed points is the empty set, a single point, a line or the whole plane.*

Proof Clearly, when I is the identity the set of fixed points is the whole plane. Assume I is not the identity, so there exists at least one point s for which $I(s) \neq s$. Our first observation is that *any* fixed point p must lie on the orthogonal bisector of s, $I(s)$: indeed $|I(s) - p| = |I(s) - I(p)| = |s - p|$, so p is equidistant from s, $I(s)$. If now I has no fixed points, or exactly one, there is nothing further to prove, so we can assume I has at least two distinct fixed points p, q: and the line joining p, q must be the orthogonal bisector of s, $I(s)$. It remains to show that any point r on that line is a fixed point. Suppose that $I(r) \neq r$. Then, replacing s by r in the above arguments, we see that p, q also lie on the orthogonal bisector of $r, I(r)$. However, that is a contradiction, since p, q, r are collinear, establishing that $I(r) = r$ as required. □

We are now in a position to write down explicit formulas for isometries, enabling us to distinguish two fundamentally different geometric types.

Lemma 6.5 *Any isometry I is given either by a unique formula $I(z) = uz + b$ or by a unique formula $I(z) = u\bar{z} + b$, for some complex numbers u, b with u unit.*

Proof Set $b = I(0)$ and define an isometry J with $J(0) = 0$ by $J(z) = I(z) - b$. Now set $u = J(1)$ and observe that u is unit, since

$$|u| = |J(1)| = |J(1) - 0| = |J(1) - J(0)| = |1 - 0| = 1.$$

Define a further isometry K by $K(z) = u^{-1}J(z)$, and note that $K(0) = 0$, $K(1) = 1$ so 0, 1 are two distinct fixed points of K. It follows from Lemma 6.4 that the set of fixed points for K is either the whole plane, or the line $y = 0$ through 0, 1. In the former case $K(z) = z$, and hence $I(z) = uz + b$, for all z. In the latter case we claim that $K(z) = \bar{z}$, and hence that $I(z) = u\bar{z} + b$, for all points z. Note first that if $K(z) = z$ then z is a fixed point for K, so lies on the line $y = 0$ and automatically satisfies $K(z) = \bar{z}$: we can therefore assume that $K(z) \neq z$. Since K is an isometry:

$$|K(z) - 0| = |K(z) - K(0)| = |z - 0|$$
$$|K(z) - 1| = |K(z) - K(1)| = |z - 1|.$$

Thus z, $K(z)$ are equidistant from 0, and from 1. It is immediate that $K(z) = \bar{z}$, as was required. (Example 1.12.) Finally, uniqueness follows from Lemma 6.2. □

Isometries of the form $I(z) = uz + b$ are said to be *direct* whilst those of the form $I(z) = u\bar{z} + b$ are *indirect*. At this stage in our development we are only concerned with direct isometries, though we will meet indirect isometries as 'reflexions' in Chapter 11. Direct isometries are also known as *congruences*. (There is a possible confusion with standard usage in school geometry where the term 'congruence' is used as a synonym for 'isometry'.) The congruences form a subgroup of the isometry group, known as the *special* Euclidean group and denoted $SE(2)$. (Exercise 6.2.1) The virtue of having explicit formulas for isometries is that we can establish their properties via simple computations, a point well illustrated by the next example. First a definition: an isometry I having exactly one fixed point c is a *rotation*: one says that c is the *centre* of the rotation, and that the rotation is *about* c. (It is a convention that the identity isometry is regarded as a rotation.)

Example 6.3 *Any direct isometry I is either a translation or a rotation.* By the above we can write $I(z) = uz + b$ for complex numbers u, b with

u unit. Translations correspond to $u = 1$: in that case there are no fixed points, unless $b = 0$ when I is the identity. However, when $u \neq 1$ fixed points are given by $I(z) = z$, i.e. $(u - 1)z = -b$ with unique solution $z = b/(1-u)$: in that case I is a rotation, and we say that it is through an angle θ when $u = e^{i\theta}$ with $0 \le \theta < 2\pi$. In particular, a direct isometry I with $I(0) = 0$ is a rotation about the origin, given by a formula $I(z) = uz$. Writing $I(z) = Z$, with $z = x + iy$, $Z = X + iY$ we derive the familiar formulas for a rotation about the origin through an angle θ, namely

$$X = x\cos\theta - y\sin\theta, \quad Y = x\sin\theta + y\cos\theta.$$

Example 6.4 By *central reflexion* in a point p we mean the planar map I defined by $I(z) = 2p - z$. This has the form $I(z) = uz + b$ with $u = -1$, $b = 2p$ so is a direct isometry. Note that I is defined by the condition that for any $z \neq p$ the point p is the mid-point of the line segment joining z, $I(z)$. There is a unique fixed point p, so I is rotation about p through an angle π, a so-called *half turn*. Central reflexion in the origin $p = 0$ is given by $I(z) = -z$.

Example 6.5 In Lemma 6.2 we showed that any isometry I can be written *uniquely* as the composite of a translation T and an isometry R with the property that $R(0) = 0$. The translation T is necessarily given by a formula $T(z) = z + b$ for some complex number b: and Lemma 6.5 shows that R is given *either* by a formula $R(z) = uz$ *or* by a formula $R(z) = u\bar{z}$, where u is some fixed unit complex number. Either way we see that R is a linear mapping, known as the *linear part* of I.

Exercises

6.2.1 Show that the congruences form a subgroup of the isometry group, and that the rotations with centre p form a subgroup of the congruence group.

6.2.2 Let p be a fixed point in the plane. Show that for each rotation R with centre the origin there is a unique rotation I_R with centre p whose linear part is R. Use this to show that the group of rotations with centre the origin is isomorphic to the group of rotations with centre p.

6.2.3 Let I be an indirect isometry given by $I(z) = u\bar{z} + b$ with u unit. Show that the equation $I(z) = z$ has a solution z if and only if $I(b) = 0$.

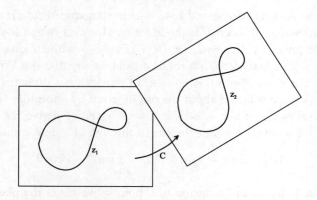

Figure 6.2. Congruent curves

6.2.4 An isometry I is of *order* 2 in the isometry group when $I^2 = 1$, $I \neq 1$. Show that any isometry I of order 2 has at least one fixed point p. (Consider the mid-point p of the line segment joining r, $I(r)$ for some r.) Use this fact to deduce that a congruence is of order 2 if and only if it is a half turn.

6.3 Congruent Curves

We are now in a position to return to the question of when two curves z_1, z_2 with the same domain I should be regarded as 'the same'. In this section (and the next) we write T_1, T_2 for the unit tangent vectors, N_1, N_2 for the unit normal vectors, s_1, s_2 for the speeds, and κ_1, κ_1 for the curvature functions. For notational efficiency we may drop the variable t. We say that z_1, z_2 are *congruent* when there exists a congruence C with the property that $z_2(t) = C(z_1(t))$ for all $t \in I$. One pictures the idea as in Figure 6.2. When C is a translation z_1, z_2 are *translationally* congruent: and when C is a rotation z_1, z_2 are *rotationally* congruent.

The relation of congruence defines an equivalence relation on the set of curves with domain a fixed open interval I. (Exercise 6.3.1.) Here is an interesting class of examples to illustrate the idea.

Example 6.6 Let p be a point in the plane. By an *equiangular spiral* with *pole* p we mean a regular curve z with the property that $z'(t) = \gamma(z(t) - p)$ for some fixed complex number γ. In particular, an equiangular spiral with pole the origin is characterized by the relation $z'(t) = \gamma z(t)$. Note that *any* equiangular spiral z is translationally congruent to an equiangular

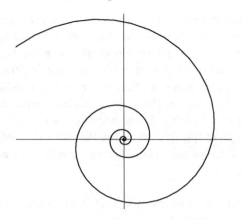

Figure 6.3. An equiangular spiral

spiral w with pole the origin, for writing $w(t) = z(t) - p$ we obtain $w'(t) = \gamma w(t)$, so w is an equiangular spiral with pole the origin. Taking moduli gives $|w'(t)| = |\gamma||w(t)|$, so the speed is a constant scalar multiple of the length of the vector $w(t)$. Since the curve is assumed to be regular, we deduce that $\gamma \neq 0$, and that the pole itself cannot lie on the spiral. The description 'equiangular' refers to the fact that the angle θ (Figure 6.3) between the tangent line at the parameter t and the line joining p to $z(t)$ must be constant, since its cosine is

$$\cos\theta = \frac{w(t)\bullet w'(t)}{|w(t)||w'(t)|} = \frac{w(t)\bullet \gamma w(t)}{|w(t)||\gamma w(t)|} = \frac{w(t)\bullet \gamma w(t)}{|w(t)||\gamma||w(t)|} = \frac{\Re(\gamma)}{|\gamma|}.$$

How do we know that equiangular spirals actually exist? One way is to write down an explicit parametrization.

Example 6.7 By a *standard* equiangular spiral we mean a curve z given by a formula $z(t) = re^{\gamma t}$, where r is a positive real number, and $\gamma = \alpha + i\beta$ with α, β real numbers, not both zero. Here $z'(t) = r\gamma e^{\gamma t} = \gamma z(t)$, so according to the above definition z is an equiangular spiral with pole the origin. Note that there are two *degenerate* cases. When $\alpha = 0$ we have $z(t) = re^{i\beta t}$ so the curve is a parametrization of the circle radius r centred at the origin, with angle $\theta = \pi/2$: and when $\beta = 0$ we have $z(t) = re^{\alpha t}$ so the curve is a parametrization of the positive x-axis, with angle $\theta = 0$.

Example 6.8 Had we not been aware of such parametrizations we could have *deduced* their existence as follows. The defining relation $z'(t) = \gamma z(t)$

is a complex linear differential equation with general solution $z(t) = z_0 e^{\gamma t}$ where z_0 is some fixed complex number. We are only interested in the case when z_0 and γ are non-zero, since otherwise the trace of the curve is a single point, and every parameter is irregular. Writing $z_0 = re^{i\phi}$ with $r > 0$ we see that our curve is obtained from the standard equiangular spiral $z(t) = re^{\gamma t}$ by rotation through an angle ϕ about the origin. Thus any equiangular spiral with pole the origin is rotationally congruent to a standard equiangular spiral. Combining this with the above remarks, we conclude that *any equiangular spiral is congruent to a standard equiangular spiral*.

The point of the next example is that it provides natural motivation for widening the concept of congruent curves.

Example 6.9 A curve w is obtained from the standard trochoid z exhibited in (3.2) by replacing the positive constant h by $-h$: thus z, w are given by

$$z(t) = (\lambda + 1)e^{it} - he^{(\lambda+1)it}, \quad w(t) = (\lambda + 1)e^{it} + he^{(\lambda+1)it}$$

where it is assumed that $\lambda \neq -1$. We claim that z, w are not congruent curves. Suppose they are, so there exist complex numbers u, b with u unit for which $w(t) = uz(t) + b$ identically. Differentiation yields the identity $w'(t) = uz'(t)$. Setting $t = 0$ we obtain $w'(0) = uz'(0)$, which simplifies to $1 + h = u(1 - h)$. It follows immediately that u is real, hence that $u = \pm 1$: either way, we have a contradiction, establishing the claim. However, a few lines of working will verify that z, w satisfy the identity $w(t + \theta) = e^{i\theta} z(t)$ where $\theta = \pi/\lambda$. Thus *although w is not congruent to z, it is congruent to a reparametrization of z.*

In view of examples such as this it is natural to widen the concept of congruence as follows. We say that two curves z_1, z_2 with domains I_1, I_2 are *equivalent* when there exist a congruence C and a change of parameter $s : I_1 \to I_2$ with everywhere positive derivative for which $z_2(s(t)) = C(z_1(t))$ for all t. Note that congruent curves are necessarily equivalent (take s to be the identity): and likewise, parametrically equivalent curves are necessarily equivalent (take C to be the identity). Thus the curves z, w of the above example are equivalent, even though they fail to be congruent. We conclude the body of this section by establishing that curvature is invariant under the relation of congruence.

Lemma 6.6 *Let z_1, z_2 be congruent regular curves. Then z_1, z_2 have the same speeds s_1, s_2 and the same curvatures κ_1, κ_2.*

Proof Since z_1, z_2 are congruent there exist a rotation R about the origin, and a fixed vector w with the property that $z_2(t) = R(z_1(t)) + w$ for all t. Differentiation with respect to t gives $z_2'(t) = R(z_1'(t))$ for all t. Since rotations preserve length (Lemma 6.3) we deduce that $s_1(t) = s_2(t)$ for all t. It follows immediately that $T_2(t) = R(T_1(t))$ and $N_2(t) = R(N_1(t))$. Moreover, differentiating the former relation with respect to t yields $T_2'(t) = R(T_1'(t))$. Now, dropping the variable t for convenience, and using the fact that rotations preserve scalar products (Lemma 6.3), we obtain the desired result via

$$\kappa_2 = \frac{T_2' \bullet N_2}{s_2} = \frac{R(T_1') \bullet R(N_1)}{s_1} = \frac{T_1' \bullet N_1}{s_1} = \kappa_1.$$

□

Exercises

6.3.1 Consider the set of curves z whose domains are a fixed open interval I. For two such curves z_1, z_2 write $z_1 \sim z_2$ to mean that z_1 is congruent to z_2. Show that the relation of congruence is an equivalence relation on the set.

6.3.2 Let b, n be non-zero real numbers. Show that the curves given by $w(t) = 2be^{it} \sin nt$ are equivalent to, but not congruent to, the rose curves $z(t) = 2be^{it} \cos nt$.

6.3.3 Let p_1, q_1, p_2, q_2 be four points. Show that the line segment joining p_1, q_1 is congruent to the line segment joining p_2, q_2 if and only if $|p_1 - q_1| = |p_2 - q_2|$.

6.4 The Uniqueness Theorem

Our object now is to prove the converse of Lemma 6.6, i.e. that if two curves z_1, z_2 have equal speeds and curvatures respectively at every point then they must be congruent. Thus, up to the relation of congruence, speed and curvature determine the curve uniquely. The main step is the following result, based entirely on standard facts from foundation year calculus.

86 Existence and Uniqueness

Lemma 6.7 *Let z_1, z_2 be curves having equal speeds s_1, s_2 and equal curvatures κ_1, κ_2. Suppose there exists a parameter $t = t_0$ for which $z_1(t_0) = z_2(t_0)$ and $T_1(t_0) = T_2(t_0)$. Then z_1, z_2 coincide.*

Proof Consider the smooth function f defined by the formula

$$f(t) = T_1(t) \bullet T_2(t) + N_1(t) \bullet N_2(t).$$

We claim that the derivative $f'(t)$ is identically zero. Indeed, suppressing the parameter t for convenience, and using the Serret–Frenet Formulas, we have

$$\begin{aligned} f'(t) &= T_1 \bullet T_2' + T_1' \bullet T_2 + N_1 \bullet N_2' + N_1' \bullet N_2 \\ &= T_1 \bullet (s_2 \kappa_2 N_2) + (s_1 \kappa_1 N_1) \bullet T_2 + N_1 \bullet (-s_2 \kappa_2 T_2) \\ &\quad + (-s_1 \kappa_1 T_1) \bullet N_2 \\ &= (s_2 \kappa_2 - s_1 \kappa_1)(T_1 \bullet N_2 - N_1 \bullet T_2) = 0 \end{aligned}$$

since $s_1 = s_2$ and $\kappa_1 = \kappa_2$ by hypothesis. It follows from calculus that f is constant. Moreover, setting $t = t_0$, and using the hypothesis that $T_1(t_0) = T_2(t_0)$, hence that $N_1(t_0) = N_2(t_0)$, we get

$$\begin{aligned} f(t_0) &= T_1(t_0) \bullet T_2(t_0) + N_1(t_0) \bullet N_2(t_0) \\ &= |T_1(t_0)|^2 + |N_1(t_0)|^2 \\ &= 1 + 1 = 2 \end{aligned}$$

so that f assumes the constant value 2. Now choose any $t \in I$. By the Cauchy–Schwarz Inequality we have

$$\begin{aligned} T_1(t) \bullet T_2(t) &\leq |T_1(t)||T_2(t)| = 1 \\ N_1(t) \bullet N_2(t) &\leq |N_1(t)||N_2(t)| = 1. \end{aligned}$$

If either of these inequalities were *strict*, the value of $f(t)$ would be < 2, which we have just shown to be impossible. Thus both these inequalities are equalities, and we have $T_1(t) \bullet T_2(t) = 1$, $N_1(t) \bullet N_2(t) = 1$ for all t. It follows that

$$\begin{aligned} |T_1 - T_2|^2 &= (T_1 - T_2) \bullet (T_1 - T_2) \\ &= T_1 \bullet T_1 - 2 T_1 \bullet T_2 + T_2 \bullet T_2 \\ &= 1 - 2 + 1 = 0 \end{aligned}$$

6.4 The Uniqueness Theorem

so $T_1 = T_2$, and hence $z_1' = z_2'$. Further, for any t we have

$$z_1(t) - z_1(t_0) = \int_{t_0}^t z_1'(x)dx = \int_{t_0}^t z_2'(x)dx = z_2(t) - z_2(t_0).$$

By hypothesis $z_1(t_0) = z_2(t_0)$, so $z_1(t) = z_2(t)$ identically. □

Theorem 6.8 *Let $z_1 : I \to \mathbb{R}^2$, $z_2 : I \to \mathbb{R}^2$ be regular curves whose speeds s_1, s_2 and curvatures κ_1, κ_2 each coincide: then z_1, z_2 are congruent.* (The Uniqueness Theorem.)

Proof Choose any fixed value $t = t_0$ of the parameter. Apply a translation to z_1 to obtain a curve z_3 with $z_3(t_0) = 0$: and likewise apply a translation to z_2 to obtain a curve z_4 with $z_4(t_0) = 0$: since translations leave speed and curvature invariant, z_3, z_4 have the same speeds and the same curvatures, and have the property that $z_3(t_0) = z_4(t_0)$. Now apply a rotation (about the origin) to z_4 to obtain a new curve z_5 with $z_3'(t_0) = z_5'(t_0)$. Since rotations also leave speed and curvature invariant we see that z_3, z_5 have the same speeds and the same curvatures, and satisfy the relations $z_3(t_0) = z_5(t_0)$, $z_3'(t_0) = z_5'(t_0)$. It follows immediately from the previous proposition that z_3, z_5 coincide. But z_1, z_3 are congruent, and likewise z_2, z_5 are congruent, so z_1, z_2 are congruent, as required. □

Here are some illustrations. In Lemmas 5.4 and 5.5 it was shown that the trace of any regular curve of constant curvature κ is either a line segment (when $\kappa = 0$) or an arc of a circle (when $\kappa \neq 0$). We can deduce these results from the Uniqueness Theorem.

Example 6.10 Consider a regular curve $z : I \to \mathbb{R}^2$ of constant zero curvature. By Lemmas 4.1 and 5.2 there is a unit speed reparametrization $w : I \to \mathbb{R}^2$, having the same trace, and constant zero curvature. The line segment $I \to \mathbb{R}^2$ defined by $t \to (t, 0)$ is a unit speed curve which also has zero curvature, so by the Uniqueness Theorem is congruent to w. Thus the trace of w, and hence of z, is a line segment.

Example 6.11 Likewise, consider a regular curve $z : I \to \mathbb{R}^2$ of constant curvature $\kappa \neq 0$. By Lemma 4.1 there is a unit speed reparametrization $w : I \to \mathbb{R}^2$, having the same trace, and the same curvature κ. The arc of a circle $I \to \mathbb{R}^2$ defined by $t \to e^{i\kappa t}/\kappa$ is a unit speed curve, also having curvature κ, so by the Uniqueness Theorem is congruent to w. Thus the trace of w, and hence of z, is an arc of a circle.

Here is another application of the Uniqueness Theorem, used as a technical simplification when studying problems in planar kinematics.

Lemma 6.9 *Let $z : I \to \mathbb{R}$ be a regular curve with trace contained in the unit circle, and let t_0 be a fixed parameter. Then there exist a unit complex number u, and a change of parameter r with $r(t_0) = 0$, for which $z(t) = ue^{ir(t)}$ for all t.*

Proof By Lemma 4.1 there exists a change of parameter $s : J \to I$ with $s(0) = t_0$ for which $w(t) = z(s(t))$ has unit speed. Our first objective is to show that w has curvature $\kappa = \pm 1$. The trace of w is contained in the unit circle (parametrically equivalent curves have the same trace) so $w \bullet w = 1$ identically. Differentiating this identity twice yields $w \bullet w' = 0$ (so w, w' are orthogonal) and $w \bullet w'' + w' \bullet w' = 0$. However, since w is of unit speed we also have the identity $w' \bullet w' = 1$, yielding $w \bullet w'' = -1$. Since iw, w' are both unit vectors orthogonal to w we must have $iw' = \pm w$. Differentiating the identity $w' \bullet w' = 1$ yields $w' \bullet w'' = 0$. It follows that w and w'' are both orthogonal to w' so are linearly dependent unit vectors with $w'' = \pm w$. In fact $w'' = -w$, else the relation $w \bullet w'' = -1$ fails. We now have $\kappa = T' \bullet N = w'' \bullet iw' = -w \bullet \pm w = \mp 1$, as required. Consider the case $\kappa = 1$. The curves $w(t)$, e^{it} (the latter restricted to J) have equal speeds and curvatures, so by the Uniqueness Theorem are congruent. That means that there exist fixed complex numbers u, b with u a unit for which $w(t) = ue^{it} + b$ for all $t \in J$. Then the identity $0 = w \bullet w'$ reads $0 = (ue^{it} + b) \bullet iue^{it} = b \bullet iue^{it}$: taking two values of t for which the corresponding vectors e^{it} are linearly independent we deduce that $b = 0$, so $w(t) = ue^{it}$ for all $t \in J$. Write $r : I \to J$ for the inverse of $s : J \to I$. Then r is a change of parameter with $r(t_0) = 0$ for which $z(t) = w(r(t)) = ue^{ir(t)}$. Finally, in the case $\kappa = -1$ replace t in the preceding argument by $-t$ to reach the same conclusion. □

7
Contact with Lines

In this chapter we will discuss the way in which curves and lines intersect, via the fundamental idea of 'contact'. The key concept is the 'multiplicity' of a root s of an equation $\phi(s) = 0$. Our starting point is to extend the Factor Theorem of elementary algebra from polynomials to smooth functions: that will provide the technical tool necessary to understand the geometry. The next step is to apply this machinery to the pencil of all lines through a given point on a curve to understand how 'contact' distinguishes the tangent line at a *regular* parameter from arbitrary lines. That leads to the more subtle question of the 'contact' of the tangent line itself with the curve. The result is a characterization of inflexional parameters in terms of 'contact', and the idea of higher inflexions. In the final section we extend this line of thought to special types of *irregular* parameters ('cusps') and establish further connexions with the curvature function.

7.1 The Factor Theorem

First we require some vocabulary, extending that familiar from the theory of polynomials in a single variable. Let $\phi : I \to \mathbb{R}$ be a smooth function, with domain an open interval I. A real number t_0 which satisfies the equation $\phi(t) = 0$ is said to be a *zero* of ϕ. One of the central facts in the elementary theory of polynomials is the Factor Theorem, namely the result that if $\phi(t)$ is a *polynomial* and t_0 is a zero of ϕ then $(t - t_0)$ is a factor of $\phi(t)$, i.e. there exists a polynomial $\chi(t)$ such that $\phi(t) = (t-t_0)\chi(t)$.

Example 7.1 By way of illustration, consider the cubic polynomial $\phi(t) = t^3 + t^2 - t - 1$. The value $t = 1$ is a zero of ϕ, so $(t-1)$ must be a factor: indeed $\phi(t) = (t-1)\chi(t)$ where $\chi(t) = (t+1)^2$.

Our next result extends the Factor Theorem from polynomial functions to *smooth* functions, using facts from elementary analysis.

Theorem 7.1 *Let $\phi : I \to \mathbb{R}$ be a smooth function, having a zero at t_0. There exists a smooth function $\chi : I \to \mathbb{R}$ with $\phi(t) = (t - t_0)\chi(t)$.* (The Factor Theorem.)

Proof Consider first the special case when $t_0 = 0$, so $\phi(0) = 0$. Note that the derivative of $\phi(\lambda t)$ with respect to λ is $t\phi'(\lambda t)$ so that

$$\phi(t) = [\phi(\lambda t)]_{\lambda=0}^{\lambda=1} = \int_0^1 t\phi'(\lambda t) d\lambda = t \int_0^1 \phi'(\lambda t) d\lambda = t\chi(t)$$

provided we define χ by the formula below. It follows from a standard calculus result (differentiation under the integral sign) that χ is smooth.

$$\chi(t) = \int_0^1 \phi'(\lambda t) d\lambda.$$

The general case is deduced as follows. Set $\phi^*(t) = \phi(t + t_0)$ so $t = 0$ is a zero of ϕ^* if and only if t_0 is a zero of ϕ. Then, by the special case, there exists a smooth function χ^* with $\phi^*(t) = t\chi^*(t)$. The general case follows on writing $t - t_0$ in place of t, and defining χ by $\chi(t) = \chi^*(t - t_0)$. □

It is worth noting that the Factor Theorem of elementary algebra is a special case of this result: indeed, when $\phi(t)$ is polynomial, automatically $\chi(t)$ is polynomial as well.

Example 7.2 The smooth function $\phi : \mathbb{R} \to \mathbb{R}$ defined by $\phi(t) = \sin t$ has a zero at $t = 0$, so by the Factor Theorem there exists a smooth function $\chi : \mathbb{R} \to \mathbb{R}$ with $\phi(t) = t\chi(t)$: indeed the proof shows that

$$\chi(t) = \int_0^1 \cos(\lambda t)\, d\lambda = \frac{\sin t}{t}$$

where the second equality assumes that $t \neq 0$. To extend the formula to the whole real line we would need to define $\chi(0) = 1$, and then prove that the result is smooth. The Factor Theorem provides a painless way of avoiding such subtleties.

7.2 Multiplicity of a Zero

The key to the material of this chapter lies in the concept of the 'multiplicity' of a solution of a smooth equation. Let s_0 be a zero of ϕ, and let

7.2 Multiplicity of a Zero

$k \geq 1$ be an integer. We say that ϕ has a *k-fold zero*, or a *zero of finite multiplicity k* at s_0, when

$$\phi(s_0) = 0, \ \phi'(s_0) = 0, \ \ldots, \ \phi^{(k-1)}(s_0) = 0, \ \phi^{(k)}(s_0) \neq 0.$$

And s_0 is a zero of *infinite multiplicity* when $\phi^{(k)}(s_0) = 0$ for all positive integers k: in particular, all the zeros of the zero function ϕ are of infinite multiplicity. The reader is warned that non-zero functions can have zeros of infinite multiplicity. Such pathologies do not appear in this book, so we will say no more about them. Further, s_0 is a *repeated zero* when it is a zero of multiplicity ≥ 2: in particular, zeros of infinite multiplicity are repeated. For reasons which will become clearer as we proceed, it is important for us to know that the concept of 'multiplicity' is invariant under changes of parameter, in the following sense.

Lemma 7.2 *Let $\phi : I \to \mathbb{R}$ be a smooth function, let $s : J \to I$ be a change of parameters, and let $\psi : J \to \mathbb{R}$ be the composite function defined by $\psi(t) = \phi(s(t))$. Then $s(t)$ is a zero of order $\geq k$ of $\phi(s)$ if and only if t is a zero of order $\geq k$ of $\psi(t)$.*

Proof First, we prove by induction that for each integer $k \geq 1$ there exist smooth functions $\sigma_{k1}(t), \ldots, \sigma_{kk}(t)$ for which

$$\psi^{(k)}(t) = \sigma_{k1}(t)\phi'(s(t)) + \cdots + \sigma_{kk}(t)\phi^{(k)}(s(t)).$$

For $k = 1$ we can take $\sigma_{11}(t) = s'(t)$: and assuming the result is true for k it follows for $(k+1)$ by differentiating and using the standard rules of elementary calculus. Thus if the first k derivatives of ϕ vanish at $s(t)$ then the first k derivatives of ψ vanish at t: in other words, if $s(t)$ is a zero of order $\geq k$ of $\phi(s)$ then t is a zero of order $\geq k$ of $\psi(s)$. The result follows on reversing the roles of ϕ, ψ and replacing s by its inverse. □

The Factor Theorem can now be stated in a sharper form, providing a compelling picture of the local behaviour of a smooth function ϕ close to a zero of given multiplicity.

Theorem 7.3 *Let $\phi : I \to \mathbb{R}$ be a smooth function, having a zero of finite multiplicity k at t_0. Then there exists a smooth function $\psi : I \to \mathbb{R}$ with $\psi(t_0) \neq 0$ for which $\phi(t) = (t - t_0)^k \psi(t)$. In particular, when k is odd $\phi(t)$ changes sign at t_0: and when k is even $\phi(t)$ has constant sign close to t_0, namely that of $\phi^{(k)}(t_0)$.*

Proof By the Factor Theorem there is a smooth function $\chi : I \to \mathbb{R}$ for which $\phi(t) = (t - t_0)\chi(t)$. We claim first that t_0 is a zero of χ of multiplicity $(k-1)$. Let $n \geq 1$ be an integer. Differentiating the relation $\phi(t) = (t - t_0)\chi(t)$ successively n times, and then setting $t = t_0$, we obtain $\phi^{(n)}(0) = n\chi^{(n-1)}(0)$: it then follows immediately from the definitions that t_0 is a zero of multiplicity k of ϕ if and only if t_0 is a zero of multiplicity $(k-1)$ of χ. The main statement now follows immediately by successive applications of the Factor Theorem. For the remainder, we have only to observe that in the relation $\phi(t) = (t - t_0)^k \psi(t)$ the sign of $\psi(t)$ is constant for t close to t_0, so that the sign of $\phi(t)$ is given by that of $(t - t_0)^k$: for k odd this changes sign at t_0, and for k even it has constant sign close to t_0. □

7.3 Contact with Lines

Let $z : I \to \mathbb{R}^2$ be a curve. In this section we shall study in some detail the local question of how a line can meet the curve at some fixed point $z(t_0)$. Let u be a unit vector, representing a direction in the plane, and let L_u be the line through $z(t_0)$ orthogonal to u, with equation $\{z - z(t_0)\} \bullet u = 0$. The parameters t for which the point $z(t)$ is a point of intersection of the curve with L_u are given by the equation $\gamma(t) = 0$, where $\gamma : I \to \mathbb{R}$ is the smooth function defined by

$$\gamma(t) = \{z(t) - z(t_0)\} \bullet u. \tag{7.1}$$

We call γ the *contact function* of z with L_u at t_0. Thus we have defined a *family* of contact functions, one for each unit vector u. The function γ vanishes if and only if $z(t)$ lies on L_u, is positive on one side of that line, and is negative on the other side. The parameter t_0 is automatically a zero of $\gamma(t)$, corresponding to the intersection of L_u at the point $z(t_0)$. The line L_u is said to have *k point contact* with the curve z at $t = t_0$ when the contact function γ has a zero of multiplicity k at t_0: and L_u has *infinite contact* with z at $t = t_0$ when γ has a zero of infinite multiplicity at t_0. For convenience we will sometimes refer to *even* (respectively *odd*) contact when γ has a zero of even (respectively odd) finite multiplicity at t_0.

Example 7.3 The eight-curve $x(t) = a\cos t$, $y(t) = a\sin t \cos t$ where $a > 0$ has the property that $z(\pi/2) = (0,0)$. We will calculate its contact at the parameter $t_0 = \pi/2$ with $y = x$, i.e. the line L_u where $u = (1,-1)$. The contact function with L_u at the origin is $\gamma(t) = a\cos t(1 - \sin t)$. A

7.3 Contact with Lines

line or two of working verifies that $\gamma'(t_0) = 0$, $\gamma''(t_0) = 0$, $\gamma'''(t_0) = -3 \neq 0$, so the line L_u has three point contact with z at t_0.

Before pursuing the question of contact with lines we need to clarify its basic invariance properties. First, it is invariant under parametric equivalence, in the following precise sense.

Lemma 7.4 *Let z, w be curves with domains I, J, parametrically equivalent via the change of parameters $s : J \to I$. Then the contact of the line L_u with z at $s(t_0)$ equals that of L_u with w at t_0.*

Proof Let $\gamma(t) = \{z(t) - z(s(t_0))\} \bullet u$ be the contact function for z with L_u at $s(t_0)$. Then the contact function for w with L_u at t_0 is

$$\delta(t) = \{w(t) - w(t_0)\} \bullet u = \{z(s(t)) - z(s(t_0))\} \bullet u = \gamma(s(t)).$$

It follows immediately from Lemma 7.2 that t_0 is a zero of δ of multiplicity k if and only if $s(t_0)$ is a zero of γ of multiplicity k. The result now follows from the definitions. □

However, contact with lines is also invariant under congruences, in the following sense.

Lemma 7.5 *Let I be an isometry, and let z, w be curves with $w(t) = I(z(t))$. Then the contact of the line L_u with z at t_0 equals that of the line $I(L_u)$ with w at t_0.*

Proof Recall first that any isometry is the composite of an isometry R with $R(0) = 0$ and a translation T. (Lemma 6.2.) Thus it suffices to prove that contact is invariant under such isometries R, and translations T. Consider first the case of an isometry R with $R(0) = 0$. The image $R(L_u)$ is a line M_v through $w_0 = R(z(t_0))$ where $v = R(u)$. Let $\gamma(t) = \{z(t) - z(t_0)\} \bullet u$ be the contact function for z with L_u at t_0. Then the contact function for w with L_u at t_0 is

$$\begin{aligned}\delta(t) &= \{w(t) - w(t_0)\} \bullet v = \{R(z(t)) - R(z(t_0))\} \bullet R(u) \\ &= \{z(t) - z(t_0)\} \bullet u = \gamma(t),\end{aligned}$$

where we use the fact that R preserve scalar products. (Lemma 6.3.) It remains to verify that contact is preserved by translations T. Suppose that $T(z) = z + b$, with b a complex number. Then $T(L_u)$ is a parallel

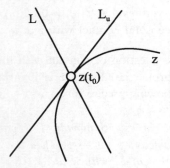

Figure 7.1. Curve crossing a non-tangent line L

line M_u, and

$$\begin{aligned}\delta(t) &= \{w(t) - w(t_0)\} \bullet u = \{T(z(t)) - T(z(t_0))\} \bullet u \\ &= \{(z(t) + b) - (z(t_0) + b)\} \bullet u \\ &= \{z(t) - z(t_0)\} \bullet u = \gamma(t).\end{aligned}$$

□

7.4 Inflexions and Undulations

We are now in a position to derive useful geometric information from the contact function. For the moment we restrict ourselves to the case when t_0 is a regular parameter. (In the next section we will extend our discussion to certain types of irregular parameter.) Our opening observation is that the tangent lines to a curve are *characterized* by having contact of order ≥ 2.

Lemma 7.6 *At a regular parameter t_0 for a curve z the tangent line is the only line through $z(t_0)$ which has contact of order ≥ 2 with the curve at $z(t_0)$. In particular, any other line L through $z(t_0)$ has contact of order 1, and the curve must cross L at that point. (Figure 7.1.)*

Proof Write $\gamma(t) = \{z(t) - z(t_0)\} \bullet u$ for the contact function. The line L_u is tangent at t_0 if and only if $\gamma'(t_0) = 0$, i.e. if and only if $z'(t) \bullet u = 0$, i.e. the vector u is parallel to the normal vector at t_0. Thus the tangent line at t_0 is the *only* line L_u through $z(t_0)$ tangent to the curve at that point. For any other line L through z_0 the contact function γ has a zero

7.4 Inflexions and Undulations

of order 1 at t_0, so by Theorem 7.3 changes sign at t_0. The result follows from previous remarks. □

Let us concentrate now on the more subtle question of how a curve z behaves relative to the tangent line itself. Thus we can assume in the following that $u = iz'(t_0)$, the normal vector at t_0, so that

$$\gamma(t) = \{z(t) - z(t_0)\} \bullet iz'(t_0).$$

The next result relates contact of order 2 to the curvature. In particular it provides a compelling interpretation for the sign of the curvature function underpinning one's mental picture of the whole subject. Before reading the proof you may like to refer back to Example 1.9 where we discussed the two sides of a line.

Lemma 7.7 *At a regular parameter t_0 a curve z has contact of order 2 with the tangent line if and only if the curvature $\kappa(t_0) \neq 0$. For all parameters t sufficiently close to t_0 the point $z(t)$ lies on the same side of the tangent line as the unit normal when $\kappa(t_0) > 0$, and on the opposite side when $\kappa(t_0) < 0$.*

Proof In view of the invariance results for contact it is no restriction to assume that z is a unit speed curve. Differentiating the contact function twice, and setting $t = t_0$, we obtain the relations $\gamma'(t_0) = 0$, $\gamma''(t_0) = z''(t_0) \bullet iz'(t_0)$. Now observe that

$$\kappa(t_0) = T'(t_0) \bullet N(t_0) = z''(t_0) \bullet iz'(t_0) = \gamma''(t_0).$$

It is immediate that we have contact of order 2 if and only if $\kappa(t_0) \neq 0$. In that case the sign of $\kappa(t_0)$ agrees with that of $\gamma''(t_0)$. However, by Theorem 7.3 the sign of $\gamma''(t_0)$ agrees with that of $\gamma(t_0)$, so $\kappa(t_0)$ has the same sign as $\gamma(t_0)$. The result is immediate from the observation that the contact function $\gamma(t)$ is positive on the side of the tangent line containing the unit normal vector, and negative on the opposite side. □

Of course, we can rephrase the first statement by saying that the contact at the parameter t has order ≥ 3 if and only if $\kappa(t) = 0$: equivalently, the contact at t has order ≥ 3 if and only if t is inflexional. Recall that the condition for a regular parameter t to be inflexional is that

$$x'(t)y''(t) - x''(t)y'(t) = 0. \tag{7.2}$$

The concept of contact allows us to explore the fine detail of inflexions. Inflexional parameters of contact 3 are *ordinary*, and those of contact ≥ 4 are *undulational*. In principle it is not difficult to locate undulational

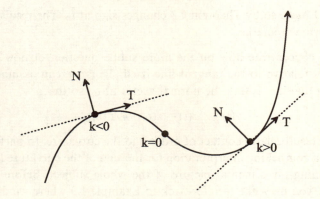

Figure 7.2. Picturing the sign of the curvature

parameters. The condition for an inflexional parameter t to be undulational is that $0 = \gamma'''(t) = z'''(t) \bullet iz'(t)$, i.e. that the vectors $z'(t)$, $z'''(t)$ should be linearly dependent, so that *in addition* to (7.2) we have

$$x'(t)y'''(t) - x'''(t)y'(t) = 0. \tag{7.3}$$

Example 7.4 Consider the graph $y = f(x)$ of a smooth function $f(x)$ viewed as the curve $x(t) = t$, $y(t) = f(t)$. As in Example 5.5 the condition for an inflexion simplifies to $f''(t) = 0$; and the conditions for an undulation simplify to $f''(t) = 0$, $f'''(t) = 0$. For instance when $f(t) = t^n$ with $n \geq 3$ these conditions reduce to $n(n-1)t^{n-2} = 0$, $n(n-1)(n-2)t^{n-3} = 0$: there is therefore a unique inflexion when $t = 0$, which is an undulation if and only if $n \geq 4$.

Lemma 7.8 *Close to an inflexional parameter t_0 having even contact the curve stays to one side of the tangent line at t_0. And at an inflexional parameter t_0 having odd order the curve crosses the tangent line at t_0: in particular, that is the case at an ordinary inflexion.* (Figure 7.3).

Proof When t_0 is an inflexion with even contact Theorem 7.3 ensures that $\gamma(t)$ has constant sign close to t_0; and when t_0 is an inflexion with odd contact $\gamma(t)$ changes sign at t_0. The result now follows exactly as in Lemma 7.6. □

Example 7.5 In Example 7.4 we saw that the curve $x(t) = t$, $y(t) = t^n$ with $n \geq 3$ has a unique inflexion when $t = 0$ which is an undulation if

7.4 Inflexions and Undulations

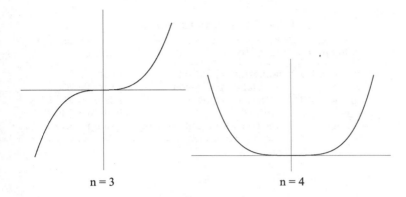

Figure 7.3. Inflexions having odd and even contact

and only if $n \geq 4$. Moreover, the contact function at $t = 0$ is $\gamma(t) = t^n$ having a zero of multiplicity n. When n is odd the contact is odd, and the graph crosses the tangent line $y = 0$: and when n is even the contact is even and the graph stays on one side of the tangent line. Figure 7.3 illustrates this result for the cases $n = 3$, $n = 4$.

Example 7.6 A *Serpentine* is an algebraic curve defined by an equation $x^2y - b^2x + a^2y = 0$ where $a, b > 0$. Its zero set is the graph of a function: indeed, solving the equation for y we find that $y = f(x)$ where

$$f(x) = \frac{b^2 x}{x^2 + a^2}.$$

According to the above, inflexional parameters are found by solving the equation $f''(x) = 0$. We leave the reader to check that that gives three inflexions at $x = 0$, $x = \sqrt{3}a$, $x = -\sqrt{3}a$. All are ordinary, since $f'''(x)$ takes the respective values $-6c$, $3c/16$, $3c/16$ where $c = a^2/b^4$, none of which vanish. (Figure 7.4.)

Example 7.7 We will investigate undulations in the family of limacons $z(t) = 2e^{it} - he^{2it}$, where $h > 0$. (Example 3.13.) Recall that the curve is regular, save in the exceptional case $h = 1$. The inflexions were determined in Example 5.9. We found that the curve has an inflexion if and only if h lies in the range $1/2 \leq h < 1$: in fact there is just one inflexion when $h = 1/2$ (the *uniflexional* case) and just two when $1/2 < h < 1$ (the *biflexional* case). The conditions for an undulation are that (7.2) and (7.3) hold simultaneously. In Example 5.9 we saw that the former condition

Table 7.1. *Inflexions on limacons*

value of h	geometry	reference
$0 < h < 1/2$	uniflexional limacon	Example 7.7
$h = 1/2$	undulational limacon	Example 7.7
$1/2 < h < 1$	biflexional limacon	Example 5.9
$h = 1$	cardioid	Example 3.13
$h > 1$	limacon with a self crossing	Example 3.14

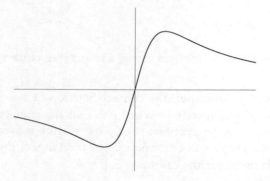

Figure 7.4. The Serpentine

reduces to $\cos t = H$ where $H = (1 + 2h^2)/3h$, whilst the latter reduces to $-12h \sin t = 0$. Clearly, these relations can only hold simultaneously when $H = 1$, and hence $h = 1/2$. For this reason the limacon with $h = 1/2$ is said to be *undulational*. Thus the geometry enables us to distinguish five classes of limacon, presented in Table 7.1.

The condition for an undulation can also be phrased purely in terms of the curvature function, providing a simple computational method for finding such points.

Lemma 7.9 *Let t be a regular parameter on a curve z. A necessary and sufficient condition for t to be undulational is that $\kappa(t) = 0$, $\kappa'(t) = 0$.*

Proof Recall that the condition for t to be inflexional is that (7.2) holds. Differentiating both sides of the formula (5.5) for the curvature, and setting $x'y'' - x''y' = 0$, we see that $\kappa' = 0$ if and only if $x'y''' - x'''y' = 0$, which is the condition (7.3) for t to be undulational. □

Exercises

7.4.1 Show that the curve $x(t) = t^2 + 1$, $y(t) = t^3 + t$ has two inflexions, both ordinary.

7.4.2 The inflexional parameters of the eight-curve $x(t) = a\cos t$, $y(t) = a\sin t \cos t$ with $a > 0$ were determined in Exercise 5.4.5. Show that they are all ordinary.

7.4.3 Show that the cycloid $x(t) = R(t - h\sin t)$, $y(t) = R(1 - h\cos t)$ with $h > 0$ discussed in Example 3.15 has inflexions if and only if $h < 1$, and that in that case there are infinitely many, all ordinary.

7.5 Cusps

In order better to understand the situation at an irregular parameter it will help to look more closely at the case of a regular parameter. The *gradient* of a curve z at a regular parameter t is defined to be the 'ratio' $y'(t) : x'(t)$.† At an irregular parameter t_0 we say that the curve has a *limiting gradient* when the gradient $y'(t) : x'(t)$ tends to a limiting ratio as $t \to t_0$. And in that case we define the *limiting tangent line* at t_0 to be the line through $z(t_0)$ having that limiting gradient.

Example 7.8 For the semicubical parabola $x(t) = t^2$, $y(t) = t^3$ the gradient at a regular parameter t is $3t/2$. The limiting gradient at the irregular parameter $t = 0$ is 0, and the limiting tangent line at that point is the line through $(0,0)$ having gradient 0, i.e. the line $y = 0$.

In general there is no reason to suppose that a limiting tangent exists at an irregular parameter. However, such examples are pathological, and unlikely to be met in the physical sciences. In any given example it is virtually certain that a limiting tangent line will exist. Here is a sufficient condition, covering every case likely to be met in practice.

Lemma 7.10 *Let z be a curve, and let t_0 be an irregular parameter with the property that some derivative of $z'(t)$ is non-zero at t_0. Then all parameters $t \neq t_0$ sufficiently close to t_0 are regular, and the limiting gradient exists at t_0.*

† Formally, we define an equivalence relation $\sim$ on the non-zero vectors by writing $(a,b) \sim (c,d)$ when there exists a real scalar $\lambda \neq 0$ with $(a,b) = \lambda(c,d)$. The equivalence class containing (a,b) is the *ratio* $a : b$. Geometrically, it is identified with the line $bx = ay$ through the origin, and arithmetically with the scalar a/b, interpreted as ∞ when $b = 0$.

Contact with Lines

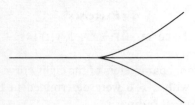

Figure 7.5. A cuspidal tangent line

Proof Let k be the least integer for which $z^{(k)}(t_0) = 0$. By Theorem 7.3 there exist smooth functions X, Y for which $x'(t) = (t - t_0)^k X(t)$, $y'(t) = (t - t_0)^k Y(t)$ and at least one of $X(t_0)$, $Y(t_0)$ is non-zero. By continuity, for $t \neq t_0$ sufficiently close to t_0 at least one of $X(t)$, $Y(t)$ is non-zero, and hence t is regular. The limiting gradient is $Y(t_0) : X(t_0)$. □

The simplest possible situation covered by this result is when $z'(t) = 0$, $z''(t) \neq 0$: a parameter t for which these relations hold is *cuspidal*, and the corresponding point $z(t)$ on the trace is a *cusp*: the vector $z''(t)$ is the *cuspidal tangent vector*, and the *cuspidal tangent line* is the limiting tangent line, i.e. the line through $z(t)$ in the direction of $z''(t)$. Note that if t is cuspidal then parameters sufficiently close to t are regular. (Exercise 7.5.8.) The next result extends Lemma 7.6 from regular parameters to cuspidal parameters.

Lemma 7.11 *At a cuspidal parameter t_0 for a curve z the cuspidal tangent line is the only line through $z(t_0)$ which has contact of order ≥ 3 with the curve at t_0. Any other line through $z(t_0)$ has contact of order 2, and the curve stays on one side of the line for parameters t sufficiently close to t_0.*

Proof With the established notation we have $\gamma(t_0) = 0$, $\gamma'(t_0) = 0$, $\gamma''(t_0) = z''(t_0) \bullet u$. We have contact of order ≥ 3 when $\gamma''(t_0) = 0$, i.e. when $z''(t_0)$ is orthogonal to u, which means that L_u is the cuspidal tangent line. For any other line γ has a zero of order 2 at t_0, so by Theorem 7.3 has constant sign close to t_0. (Figure 7.5.) □

Example 7.9 In Example 7.8 we saw that for the semicubical parabola $x(t) = t^2$, $y(t) = t^3$ the cuspidal tangent line at the unique irregular parameter $t = 0$ is the line $y = 0$. Arbitrarily close to the origin there are parts of the trace on either side of the cuspidal tangent. However, for

7.5 Cusps

any other line through the origin the trace stays on one side, provided t is sufficiently small. Figure 7.5 illustrates this for the line $y = x$.

Let t_0 be a cuspidal parameter for a curve z. To study the contact of z with the cuspidal tangent we take the vector u in the contact function to be $u = iz''(t_0)$, so

$$\gamma(t) = \{z(t) - z(t_0)\} \bullet iz''(t_0).$$

This vanishes if and only if the point $z(t)$ lies on the cuspidal tangent line, is positive on one side of that line, and negative on the other. Differentiating the formula for $\gamma(t)$ successively, and setting $t = t_0$ in the results, we obtain

$$\gamma(t_0) = 0, \quad \gamma'(t_0) = 0, \quad \gamma''(t_0) = 0, \quad \gamma'''(t_0) = z'''(t_0) \bullet iz''(t_0)$$

so γ has a zero of order ≥ 3 at t_0. A cuspidal parameter t_0 is said to have *order k* when γ has a zero of order $(k+2)$ at t_0. Cusps of order 1 are said to be *ordinary*, and those of order ≥ 2 are *higher*. Evidently, a cusp is ordinary if and only if the vectors $z''(t_0)$, $z'''(t_0)$ are linearly independent.

Example 7.10 The model for an ordinary cusp is the semicubical parabola $z(t) = (t^2, t^3)$ illustrated in Figure 2.10. The parameter $t = 0$ is an ordinary cusp of that curve since $z'(0) = (0,0)$, $z''(0) = (2,0)$, $z'''(0) = (0,6)$ and the last two vectors are linearly independent. Note that the cuspidal tangent is the line through the origin in the direction of $z''(0) = (2,0)$, i.e. the *x*-axis.

Example 7.11 The limacons $z(t) = 2e^{it} - he^{2it}$ of Example 3.13 provide us with another instance of an ordinary cusp. We saw that z has an irregular parameter if and only if $h = 1$ (the cardioid) and that in that case $t = 0$ is the only irregular parameter in the range $0 \leq t < 2\pi$. The parameter 0 is a cusp since $z''(0) = (-2,0)$ is non-zero, and it is ordinary since $z'''(0) = (0,-6)$ is not a scalar multiple of $z''(0)$.

Example 7.12 Consider the curve $z(t) = (t^2 + t^3, t^4)$ illustrated in Figure 7.6. The only irregular parameter is $t = 0$, which is a higher cusp since $z'(0) = (0,0)$, $z''(0) = (2,0)$, $z'''(0) = (6,0)$ and the last two vectors are linearly dependent. For this example $\gamma(t) = 2t^4$ and the parameter $t = 0$ is a zero of multiplicity 4. By Lemma 7.12 below, the curve must stay on one side of the cuspidal tangent for t close to 0, as indeed it does.

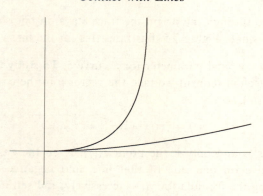

Figure 7.6. Example of a higher cusp

Lemma 7.12 *At a cuspidal parameter t_0 of odd order the curve crosses the cuspidal tangent line at t_0: in particular, this is the case at an ordinary cusp. Close to a cuspidal parameter t_0 of even order the curve stays on one side of the cuspidal tangent line.* (Figure 7.6.)

The proof of Lemma 7.12 follows immediately from Theorem 7.3. Ordinary cusps have an interesting property not enjoyed by cusps in general. By way of background, recall that in Example 5.10 we gave three examples of curves each with a cusp t_0 illustrating that the curvatures at nearby parameters t could exhibit widely differing behaviours as $t \to t_0$ through regular parameters t. The final result of this section is that for ordinary cusps the curvature tends to an infinite limit. The reader is encouraged to compare this theoretical fact with a computer generated illustration of an ordinary cusp (such as Figure 2.10) in order better to appreciate the extraordinary visual subtlety of the curvature function.

Lemma 7.13 *Let t_0 be an ordinary cusp on a curve z. Then the curvature $\kappa(t) \to \infty$ as $t \to t_0$.*

Proof As usual we write $x(t)$, $y(t)$ for the components of $z(t)$. We use the formula (5.5) for the curvature. According to that formula we can write $\kappa(t) = f(t)/g(t)^{3/2}$ where the smooth functions f, g are defined by

$$f(t) = x'(t)y''(t) - x''(t)y'(t), \quad g(t) = x'(t)^2 + y'(t)^2.$$

Note that $z'(t_0) = 0$ since t_0 is an irregular parameter, that $z''(t_0) \neq 0$ since t_0 is a cusp, and that $z''(t_0)$, $z'''(t_0)$ are linearly independent since the cusp is ordinary. Using that information we see that

$$f(t_0) = g(t_0) = 0, \quad f'(t_0) = g'(t_0) = 0, \quad f''(t_0) \neq 0, \quad g''(t_0) > 0.$$

The Factor Theorem ensures that there exist smooth functions F, G with $F(t_0) \neq 0$, $G(t_0) > 0$ for which

$$f(t) = (t - t_0)^2 F(t), \quad g(t) = (t - t_0)^2 G(t).$$

It follows that for $t \neq t_0$ the curvature κ is given by a formula of the following shape, which clearly $\to \infty$ as $t \to t_0$:

$$\kappa(t) = \frac{F(t)}{(t - t_0) G(t)^{3/2}}.$$

□

Example 7.13 The parameter $t = 0$ is an ordinary cusp of the semicubical parabola $z(t) = (t^2, t^3)$ with cuspidal tangent the x-axis. (Example 7.8.) The functions f, g of the above lemma are $f(t) = 6t^2$, $g(t) = t^2(4 + 9t^2)$ so one can take $F(t) = 6$, $G(t) = 4 + 9t^2$. And the curvature has the form $\kappa(t) = 3/4t + \cdots$ so certainly $\to \infty$ as $t \to 0$.

Exercises

7.5.1 In each of the following cases find all the cuspidal parameters, the cuspidal tangent lines at those parameters, and the contact of the cuspidal tangent lines with the given curve.

(i) $x(t) = t^2$, $y(t) = t^5$
(ii) $x(t) = t^2$, $y(t) = t^4 + t^5$
(iii) $x(t) = t^2 - t^3$, $y(t) = t^2 - t^4$
(iv) $x(t) = 3t^2 + 2t^3$, $y(t) = 2t^2 - t^4$
(v) $x(t) = 5t^2 - 2t^5$, $y(t) = 2t^2 - t^4$.

7.5.2 The piriform was introduced in Example 2.16 as the curve defined by the following formulas, where a, b are positive constants:

$$x(t) = a(1 + \cos t), \quad y(t) = b \sin t \, (1 + \cos t).$$

Show that the irregular parameters of the piriform are necessarily ordinary cusps.

7.5.3 Consider the trochoid z exhibited by the formula (3.2). It is assumed that $\lambda \neq -1$. In Example 3.10 it was shown that z has an irregular parameter if and only if $h = 1$. In that case, show that an irregular parameter t is necessarily a cusp. Further, show that t is ordinary if and only if $\lambda \neq -2$.

7.5.4 Show that the irregular parameters for the cycloid defined by the following formulas are necessarily ordinary cusps.

$$x(t) = R(t - h\sin t), \quad y(t) = R(1 - h\cos t).$$

7.5.5 Show that the unique irregular parameter $t = 0$ for the tractrix $x(t) = t - \tanh t$, $y(t) = \operatorname{sech} t$ is an ordinary cusp. (Example 2.22.)

7.5.6 Show that the unique irregular parameter for the cissoid of Diocles defined by the following formulas is an ordinary cusp.

$$x(t) = \frac{2at^2}{1+t^2}, \quad y(t) = \frac{2at^3}{1+t^2}.$$

7.5.7 The Lissajous figures were introduced in Example 4.11 as the curves defined by the following formulas. Show that an irregular parameter is necessarily a higher cusp.

$$x(t) = a_1 \sin(\omega_1 t + \phi_1), \quad y(t) = a_2 \sin(\omega_2 t + \phi_2).$$

7.5.8 Let t_0 be a cusp for a curve z. Show that all parameters $t \neq t_0$ sufficiently close to t_0 are regular. (Apply the Factor Theorem to the components of z'.)

7.5.9 Let t_0 be a higher cusp on a curve z. Modify the proof of Lemma 7.13 to show that the curvature tends to a *finite* limit as $t \to t_0$. Further, show that the limit is non-zero if and only if the vectors $z''(t_0)$, $z''''(t_0)$ are linearly independent.

8
Contact with Circles

In the previous chapter we were able to gain some insight into the local behaviour of a curve z by studying its contact with lines. In this chapter we extend that philosophy by studying its contact with circles, building on the fruitful concept of 'multiplicity' introduced in the previous chapter. Section 8.1 achieves the extension, via explicitly defined contact functions. There are very clear analogies. For instance, just as we have a unique line having at least two point contact with z at a regular parameter t, so we have a unique circle (the 'circle of curvature' at t) having at least three point contact with z at t, at least provided t is not inflexional. The locus of centres for the 'circles of curvature' gives rise to a new curve known as the 'evolute' of z, providing the subject matter for Section 8.2: it will turn out that the evolute construction can be reversed via the notion of an 'involute'. Evolutes play an important role in understanding the geometry of curves, and we will devote time to alternative descriptions. Thus in Section 8.3 evolutes are described dynamically, as the locus of irregular points for the curves 'parallel' to z. Later (Chapter 10) we will meet a third description of the evolute as the 'envelope' of the normal lines, providing the key to the crucial role played by evolutes (Chapter 12) in understanding the idea of a 'caustic'.

8.1 Contact Functions

Let t_0 be a regular parameter for a curve $z : I \to \mathbb{R}^2$. A general circle C with centre c and radius r has an equation $|z - c|^2 = r^2$; we will assume that C passes through $z(t_0)$ so that $|z(t_0) - c|^2 = r^2$. By analogy with Section 7.3 we define the *contact function* of z with C at t_0 to be the smooth function $\gamma : I \to \mathbb{R}$ given by

$$\gamma(t) = |z(t) - c|^2 - |z(t_0) - c|^2. \tag{8.1}$$

The contact function vanishes when $z(t)$ lies on C, is positive when $z(t)$ lies outside C, and is negative when $z(t)$ lies inside C. The parameter t_0 is automatically a zero of $\gamma(t)$, corresponding to the intersection with C at the point $z(t_0)$. The circle C is said to have *k point contact* with the curve z at $t = t_0$ when γ has a zero of multiplicity k at t_0: and C is *tangent* to z at t_0 when it has at least two point contact at t_0.

Example 8.1 Recall that the trace of a rose curve $z(t) = 2be^{it}\cos nt$ (where b, n are positive) lies inside the circle C of radius $2b$ centred at the origin. (Example 2.7.) The curve intersects that circle at parameters t_0 for which $\cos nt_0 = \pm 1$. The contact function for z with C at such a parameter t_0 is

$$\gamma(t) = |z(t)|^2 - |z(t_0)|^2 = 4b^2(\cos^2 nt - \cos^2 nt_0).$$

Differentiating twice, and setting $t = t_0$, we find that $\gamma'(t_0) = 0$, $\gamma''(t_0) \neq 0$. Thus we have two point contact, and the rose curve is tangent to C at all the points where it meets it. That is consistent with the illustrations for the values $n = 2, 3, \ldots$ in Figure 2.5.

For convenience we refer to *even* (respectively *odd*) contact when γ has a zero of even (respectively odd) multiplicity at t_0. By Theorem 7.3 we see that for odd contact γ changes sign at t_0, so the curve crosses the circle at that parameter; and for even contact γ has constant sign for t close to t_0, so the curve stays on one side of the circle for t close to t_0.

Lemma 8.1 *A circle C is tangent to z at a regular parameter t_0 if and only if its centre c lies on the normal line to z at t_0. The curve crosses any circle C not tangent to z at t_0.*

Proof According to the definition C is tangent to the curve at t_0 if and only if $\gamma'(t_0) = 0$. Differentiating the relation (8.1) gives $\gamma'(t) = 2(z(t) - c) \bullet z'(t)$, so the required condition is that $(z(t_0) - c) \bullet z'(t_0) = 0$, i.e. that $z(t_0) - c$ is orthogonal to the tangent vector $z'(t_0)$, i.e. if and only if the centre c lies on the normal line at t_0. When z fails to be tangent to C, the contact is of order 1, and the second statement is immediate from the above remarks. □

8.1 Contact Functions

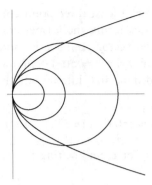

Figure 8.1. Circles tangent at the vertex of a parabola

Example 8.2 Let $c = (\alpha, \beta)$ be a point distinct from the origin, and let C be the circle centre c passing through the origin. Consider the contact of the parabola $x(t) = at^2$, $y(t) = 2at$ with C at the parameter $t = 0$. In that case the contact function is

$$\begin{aligned}\gamma(t) &= \{(x(t)-\alpha)^2 + (y(t)-\beta)^2\} - \{(x(0)-\alpha)^2 + (y(0)-\beta)^2\} \\ &= -4a\beta t + 2a(2a-\alpha)t^2 + a^2 t^4.\end{aligned}$$

The parameter $t = 0$ is always a zero of $\gamma(t)$, whose multiplicity depends on c. For $\beta \neq 0$ the multiplicity is 1; for $\beta = 0$, $\alpha \neq 2a$ the multiplicity is 2 and for $\beta = 0$, $\alpha = 2a$ the multiplicity jumps to 4. Thus C has one, two or four point contact with the parabola at $t = 0$. In particular we have tangency if and only if the centre c lies on the normal line $y = 0$ to the parabola at $t = 0$, verifying the prediction of Lemma 8.1: and in that case, provided we stay close to the origin, the parabola is *outside* C for $\alpha < 2a$, and *inside* for $\alpha \geq 2a$. In Figure 8.1 the largest circle is outside the parabola close to the origin, the smallest circle is inside, and the intermediate circle is the transitional circle of radius $2a$.

The next step is to determine when the contact is at least three point. The answer is provided by the next result.

Lemma 8.2 *At a non-inflexional regular parameter t_0 of a curve z there is exactly one circle C having at least three point contact with z at t_0, with radius $\rho = 1/|\kappa(t_0)|$: at an inflexional parameter t_0 there is no such circle.*

Contact with Circles

Proof The conditions for at least three point contact are that $\gamma'(t_0) = 0$, $\gamma''(t_0) = 0$. By the previous result, the former condition means that C is tangent to z at t_0, so the vector $z(t_0) - c$ is parallel to the unit normal vector $N(t_0)$, and we can write $c = z(t_0) + \rho N(t_0)$ for some scalar ρ. To interpret the latter condition, first differentiate the relation (8.1) twice to obtain

$$\frac{1}{2}\gamma''(t) = (z(t) - c) \bullet z''(t) + z'(t) \bullet z'(t). \tag{8.2}$$

Taking the above choice for c, and setting $t = t_0$, we see that the scalar ρ has to satisfy

$$-\rho N(t_0) \bullet z''(t_0) + z'(t_0) \bullet z'(t_0) = 0.$$

When t_0 is an inflexional parameter $N(t_0) \bullet z''(t_0) = 0$ and the equation has no solution for the scalar ρ. In other words at an inflexion there is no circle having at least three point contact with the curve. However, at a non-inflexional point there is a unique solution for ρ; using the formula for κ given in Section 5.2 it is

$$\rho = \rho(t_0) = \frac{z'(t_0) \bullet z'(t_0)}{N(t_0) \bullet z''(t_0)} = \frac{1}{\kappa(t_0)}.$$

□

Before leaving generalities about contact with circles we should point out that the reasoning used to prove Lemma 7.4 establishes the following analogue for contact with circles.

Lemma 8.3 *The order of contact of a curve with a circle is invariant under parametric equivalence, in the following sense. Let z and w be parametrically equivalent curves with domains I, J via the change of parameter $s : J \to I$. Then the contact of the circle C with w at t equals that of C with z at $s(t)$.*

Likewise the proof of Lemma 7.5 is easily adapted to prove that contact with circles is likewise invariant under isometries, in the following sense.

Lemma 8.4 *Let I be an isometry, and let z, w be curves with $w(t) = I(z(t))$. Then the contact function for the circle C with z at t_0 coincides with that for the circle $D = I(C)$ with w at t_0: in particular, the orders of contact are equal.*

Proof Let c be the centre of C. Then D is the circle centre $d = I(c)$ having the same radius. (Exercise 6.1.3.) Let γ, δ be the contact functions for z, w with C, D at t_0: then

$$\begin{aligned}\delta(t) &= |w(t) - d|^2 - |w(t_0) - d|^2 \\ &= |I(z(t)) - I(c)|^2 - |I(z(t_0)) - I(c)|^2 \\ &= |z(t) - c|^2 - |z(t_0) - c|^2 = \gamma(t).\end{aligned}$$

$\square$

Exercises

8.1.1 Find real numbers a, b for which the graph $y = x^2 + ax + b$ is tangent to the circle $x^2 + y^2 = 2$ at the point $(1, 1)$.

8.1.2 Let C be a circle, and let z be a curve whose trace is completely contained in the closed disc bounded by C. Show that z is tangent to C at any parameter t for which $z(t)$ lies on C. (Example 8.1 illustrated this result for the rose curves by direct computation.)

8.2 Evolutes

Let t be a non-inflexional regular parameter on a curve z. The scalar $\rho(t) = 1/|\kappa(t)|$ is called the *radius of curvature* at t. The unique circle of radius ρ having at least three point contact with the curve is called the *circle of curvature* at t, and its centre $z_*(t)$ defined by

$$z_*(t) = z(t) + \frac{N(t)}{\kappa(t)} \tag{8.3}$$

is the *centre of curvature* at t. Note that the circle of curvature has the same curvature at t_0 as the curve itself (up to sign). Provided z has no inflexions the resulting curve z_* is the *evolute* of z. Thus the trace of the evolute can be described as the locus of centres of curvature. With this notation, we can write down the equation of the circle of curvature at a parameter t_0: it is the unique circle through $z(t_0)$ with centre the point $z_*(t_0)$, so is given by

$$|z - z_*(t_0)|^2 = |z(t_0) - z_*(t_0)|^2. \tag{8.4}$$

It helps one's intuitive picture to regard the situation at an inflexional parameter t as a limiting case of that at a non-inflexional parameter t'. As $t' \to t$ so the radius of the circle of curvature at t' tends to infinity, so that in the limit one can think of the circle of curvature as a line having at least three point contact with the curve at t, namely the unique inflexional tangent at t.

Example 8.3 According to Lemma 5.5 any regular parametrization z of an arc A of a circle C of radius $r > 0$ has curvature of constant absolute value $|\kappa| = 1/r$. At any parameter t the circle C itself has at least three point contact (in fact infinite contact) with A, so is the circle of curvature, and the centre of curvature is the centre of C. It follows that the evolute of the arc has trace a single point, namely the centre of C.

The point of the next example is to spell out the natural invariance properties of evolutes.

Example 8.4 The evolute is invariant under congruence, in the following precise sense. Let z, w be congruent regular curves, so there exists a congruence C for which $w(t) = C(z(t))$ for all t: then the evolutes z_*, w_* are likewise congruent, indeed $w_*(t) = C(z_*(t))$ for all t. That can be deduced either from the fact that contact with circles is invariant under congruence (Lemma 8.4) or by a computation based on (8.3). Further, the evolute is invariant under parametric equivalence. Let z, w be parametrically equivalent curves, under the change of parameters s: then the evolutes z_*, w_* are likewise parametrically equivalent, under the same change of parameters s. (Exercise 8.2.11.) Combining these facts we see that evolutes are invariant under the concept of equivalence: if two curves are equivalent, then so too are their evolutes.

In practice evolutes are most readily calculated via formulas. A minor calculation using (8.3) shows that the components x_*, y_* of an evolute z_*

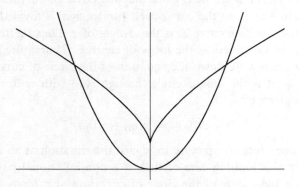

Figure 8.2. A parabola and its evolute

8.2 Evolutes

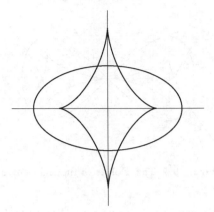

Figure 8.3. An ellipse and its evolute

are given by the following relations, where for notational economy we suppress the parameter t:

$$x_* = x - \left(\frac{x'^2 + y'^2}{x'y'' - x''y'}\right) y', \quad y_* = y + \left(\frac{x'^2 + y'^2}{x'y'' - x''y'}\right) x'.$$

Example 8.5 The speed s and curvature κ of the parabola $x(t) = at^2$, $y(t) = 2at$ with $a > 0$ are given by

$$s(t) = |z'(t)| = 2a(1 + t^2)^{1/2}, \quad \kappa(t) = \frac{-1}{2a(1 + t^2)^{3/2}}.$$

Clearly, the parabola has no inflexions. Substituting in the formula for z_* we see that $x_*(t) = 2a + 3at^2$, $y_*(t) = -2at^3$. Thus the evolute is the composite of the semicubical parabola $x(t) = t^2$, $y(t) = t^3$ with the affine mapping $X = 3ax + 2a$, $Y = -2ay$. It has only one irregular point, namely an ordinary cusp at $t = 0$ corresponding to the vertex of the parabola. (Figure 8.2.) The equation of the circle of curvature at any parameter t_0 is determined by (8.4): for instance, at $t = 0$ the circle of curvature is $x^2 + y^2 = 4ax$, the circle of radius $2a$ centre $(2a, 0)$.

Example 8.6 The ellipse $x(t) = a \cos t$, $y(t) = b \sin t$ with $0 < b < a$ has no inflexions. The formulas for the centre of curvature yield the evolute in the form

$$x_*(t) = \left(\frac{a^2 - b^2}{a}\right) \cos^3 t, \quad y_*(t) = \left(\frac{b^2 - a^2}{b}\right) \sin^3 t.$$

Figure 8.4. The evolute of the eight-curve

The evolute is obtained from the astroid (Example 4.3) by scalings of the variables. Note that the four cusps on the evolute arise from the four vertices on the ellipse.

It is not chance that vertices on the parabola and ellipse give rise to cusps on the evolute. In Chapter 9 we will introduce a general concept of 'vertex' for curves, and show that such points always give rise to cusps on the evolute. (Figure 8.3.)

Example 8.7 The evolute fails to be defined at inflexional parameters. Indeed, looking at (8.3) we see that at inflexional parameters the evolute will 'go to infinity'. A good example is provided by the eight-curve $x(t) = a\cos t$, $y(t) = a\sin t\cos t$ with $a > 0$. In the interval $0 \le t < 2\pi$ there are two inflexional parameters $t = \pi/2$, $t = 3\pi/2$ corresponding to the self crossing at the origin. Figure 8.4 illustrates that at these parameters the evolute 'goes to infinity'.

Example 8.8 The semicubical parabola z defined by $x(t) = t^2$, $y(t) = t^3$ has just one irregular parameter, namely an ordinary cusp at $t = 0$, and is easily checked to have no inflexional parameters. Thus the evolute is defined on the intervals $t > 0$ and $t < 0$ but not at $t = 0$. It is therefore somewhat surprising to find that the components of the evolute, given by the following formulas, make perfect sense when $t = 0$, defining a curve passing through $z(0)$.

$$x_*(t) = -\frac{t^2}{2}(2 + 9t^2), \qquad y_*(t) = -\frac{2t}{3}(3t^2 + 2).$$

There appears to be something amiss. However, all is well, we just have to look more carefully at (8.3) to understand what is happening. Clearly

8.2 Evolutes

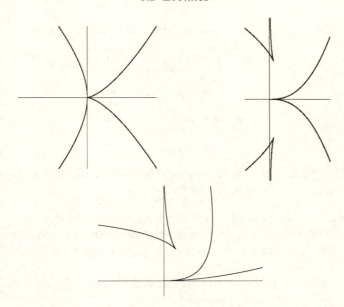

Figure 8.5. Three cusps and their evolutes

as $t \to 0$, the curvature $\kappa(t) \to \infty$, and hence $z_*(t) \to (0,0)$. Thus we expect the formula for the evolute to be defined at $t = 0$.

This example illustrates a general point, namely that *at an ordinary cuspidal parameter t_0 of a curve z the evolute will 'pass through' the corresponding point $z(t_0)$*, in the sense that $z_*(t) \to z(t_0)$ as $t \to t_0$. That follows immediately from the observation of Lemma 7.13, that at an ordinary cusp the curvature tends to an infinite limit.

Example 8.9 At this point it is instructive to look again at Example 5.10 from the 'evolute' point of view. In that example we considered the three curves z_1, z_2, z_3 defined below.

$$z_1(t) = (t^2, t^3), \quad z_2(t) = (t^2, t^5), \quad z_3(t) = (t^2 + t^3, t^4).$$

The point of that example is that although all three curves have exactly one irregular parameter, giving rise to a cusp at the origin, the limiting behaviours of the curvatures κ_1, κ_2, κ_3 are quite different. Indeed, as $t \to 0$ so $\kappa_1 \to \infty$, $\kappa_2 \to 0$ and $\kappa_3 \to 2$. Figure 8.5 illustrates the three cusps of Figure 5.4 together with their evolutes. The point to make is

Table 8.1. *Evolutes of some epicycloids and hypocycloids*

λ	name	τ	Λ
1	cardioid	π	1/3
2	nephroid	$\pi/2$	1/2
-3	deltoid	$-\pi/3$	3
-4	astroid	$-\pi/4$	2

that although the cusps look similar, their evolutes display quite vividly the differing behaviours of the curvature.

Here is another example where the irregular parameters are ordinary cusps, so that although at such parameters the evolute is not defined, the *formula* for the evolute makes sense, and indeed defines a smooth curve.

Example 8.10 In Example 3.11 we determined the irregular parameters for the epicycloids and hypocycloids

$$z(t) = (\lambda + 1)e^{it} - e^{i(\lambda+1)t}$$

where $\lambda \neq -1$. And in Example 5.8 it was established that the curve has no inflexions provided $\lambda \neq -2$. The reader is left to verify that when $\lambda \neq -2$ the evolute is given by

$$z_*(t) = \Lambda \left\{ (\lambda + 1)e^{it} + e^{i(\lambda+1)t} \right\} = \Lambda e^{-i\tau} z(t + \tau)$$

where $\Lambda = \lambda/(\lambda + 2)$, $\tau = \pi/\lambda$. (Exercise 8.2.8.) The curve $z(t + \tau)$ is a reparametrization of $z(t)$: multiplication by $\Lambda e^{i\tau}$ represents rotation about the origin through an angle τ, followed by a scaling by Λ. Table 8.1 presents four special cases of this construction. The curves and their evolutes are illustrated in Figure 8.6.

It is interesting that the evolute construction can be reversed, in the sense of the following result.

Lemma 8.5 *Let z be a regular curve, for which the involute z^* starting at t_0 is regular: then the evolute of z^* is z.*

Proof The unit tangent vector, unit normal vector, speed and curvature associated to z are written as T, N, s, κ, and those associated to z^* as T^*, N^*, s^*, κ^*. It is no restriction to assume that z has unit speed, so

8.2 Evolutes

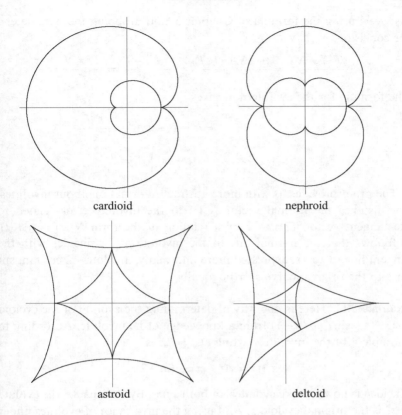

Figure 8.6. Evolutes of some epicycloids and hypocycloids

$s = 1$, and the involute is given by

$$z^*(t) = z(t) - (t - t_0)z'(t) = z(t) - (t - t_0)T(t).$$

Differentiating with respect to t, and using the Serret–Frenet Formulas, we obtain

$$(z^*)'(t) = -(t - t_0)T'(t) = -(t - t_0)\kappa(t)N(t).$$

Taking moduli of both sides of the displayed relation we see that

$$s^*(t) = \epsilon(t - t_0)\kappa(t)$$

where $\epsilon = 1$ when $(t - t_0)$, $\kappa(t)$ have opposite signs, and $\epsilon = -1$ when they have the same sign. It follows that $T^*(t) = -\epsilon N(t)$, and hence that $N^*(t) = \epsilon T(t)$. The relation between the curvatures κ, κ^* is now easy to discover: using the formula of Chapter 5, and dropping the parameter t for concision, we have

$$\kappa^* = \frac{(T^*)' \bullet N^*}{s^*} = \frac{(-\epsilon N') \bullet (\epsilon T)}{s^*} = -\frac{N' \bullet T}{s^*} = \frac{T' \bullet N}{s^*} = \frac{\kappa}{s^*}.$$

The formula for the evolute now gives

$$\begin{aligned}(z^*)_* &= z^* + \frac{N^*}{\kappa^*} = z - (t - t_0)z' + \frac{\epsilon s^* T}{\kappa} \\ &= z + \left\{ \frac{\epsilon s^*}{\kappa} - (t - t_0) \right\} T = z.\end{aligned}$$ □

The proof provides us with more abstract information about involutes. For instance the normal vector $N^*(t)$ to the involute z^* is related to the tangent vector $T(t)$ to z by a relation of the form $N^*(t) = \pm T(t)$. It follows that the normal line to the involute at t coincides with the tangent line to z at t: expressed more informally, involutes z^* cut tangent lines to the original curve z orthogonally.

Example 8.11 Here is one way of determining the evolute of the cycloid $z(t) = (t - \sin t, \cos t - 1)$ from a knowledge of its involute. According to Example 4.14 the involute starting at $t = 2\pi$ is

$$z^*(t) = z(t - \pi) + (\pi, -2).$$

The idea is now to 'take evolutes' of both sides. By Lemma 8.5 the evolute of z^* is the original cycloid z. And using the invariance of evolutes under congruence we deduce that

$$z(t) = z_*(t - \pi) + (\pi, -2).$$

Replacing the parameter t by $t + \pi$ we see that the evolute is given by

$$z_*(t) = z(t + \pi) - (\pi, -2).$$

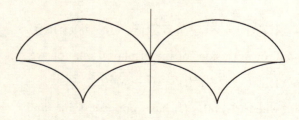

Figure 8.7. A cycloid and its evolute

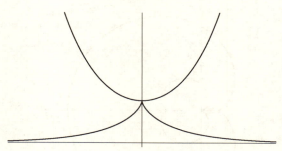

Figure 8.8. The tractrix and its evolute

It is an easy matter to verify the conclusion of this example by direct calculation. (Exercise 8.2.6.) It follows that *the evolute of a cycloid is congruent to a reparametrization of the original cycloid*, so is an equivalent curve. Both curves are illustrated in Figure 8.7.

Example 8.12 In Exercise 4.4.2 we saw that the involute of the catenary $x(t) = t$, $y(t) = \cosh t$ starting at $t = 0$ is the tractrix of Example 2.22. It follows immediately from Lemma 8.5 that the evolute of the tractrix is the catenary, a conclusion that can be verified by direct calculation. (Exercise 8.2.5.) Note incidentally that the tractrix has a simple cusp at the parameter $t = 0$, corresponding to the point $(1, 0)$ on the trace of the catenary. (Exercise 7.5.5.) Thus we have yet another illustration of the fact that the evolute of a curve 'passes through' any simple cusp. Both curves are illustrated in Figure 8.8.

Example 8.13 The concept of 'evolute' sometimes relates special curves which at first sight are unrelated. A good example is provided by Cayley's sextic $x(t) = \cos^3 t \cos 3t$, $y(t) = \cos^3 t \sin 3t$. We leave the reader to check that its evolute is a nephroid. (Exercise 8.2.10.)

Exercises

8.2.1 Find a formula for the radius of curvature ρ of the catenary $x(t) = t$, $y = \cosh t$. Let $P = (x(t), y(t))$, and let Q be the point where the normal line at t meets the x-axis. Show that the distance $PQ = \rho$.

8.2.2 Find the equation of the circle of curvature at the parameter t for the cardioid $x(t) = 2\cos t - \cos 2t$, $y(t) = 2\sin t - \sin 2t$.

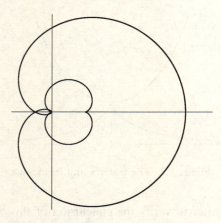

Figure 8.9. Cayley's sextic and its evolute

8.2.3 Find the equation of the circle of curvature at the parameter t for the cissoid of Diocles, given by
$$x(t) = \frac{2at^2}{1+t^2}, \quad y(t) = \frac{2at^3}{1+t^2}.$$

8.2.4 Find a formula for the evolute of the positive branch $x(t) = a\cosh t$, $y(t) = b\sinh t$ of the standard hyperbola.

8.2.5 Show that the evolute of the catenary $x(t) = t$, $y(t) = \cosh t$ is the curve $x_*(t) = t - \sinh t \cosh t$, $y_*(t) = 2\cosh t$. Find the centre of the circle of curvature at the parameter $t = 0$.

8.2.6 Verify by direct calculation that the evolute of the cycloid defined by $z(t) = (t - \sin t, \cos t - 1)$ is given by the formula
$$z_*(t) = z(t + \pi) - (\pi, -2).$$

8.2.7 Show that the evolute of the ellipse $x(t) = a\cos t$, $y(t) = b\sin t$ has no inflexions. (Example 8.6.)

8.2.8 Verify the formula given in Example 8.10 for the evolute of the epicycloid or hypocycloid
$$z(t) = (\lambda + 1)e^{it} - e^{i(\lambda+1)t}.$$

8.2.9 Let z be a standard equiangular spiral $z(t) = re^{\gamma t}$, where $\gamma = \alpha + i\beta$ is a complex number with α, β non-zero real numbers. Show that the evolute z_* is congruent to z. Deduce that the centre of curvature at t is the point where the line through the origin orthogonal to $z(t)$ meets the normal line at t.

8.2.10 Show that the evolute of Cayley's sextic $x(t) = \cos^3 t \cos 3t$, $y(t) = \cos^3 t \sin 3t$ is a nephroid.

8.2.11 Show that the evolute of a curve is invariant under parametric equivalence, in the following sense. Let $z : I \to \mathbb{R}^2$ and $w : J \to \mathbb{R}^2$ be parametrically equivalent curves, via the change of parameters $s : J \to I$. Then $z_*(t) = w_*(s(t))$.

8.2.12 Let z, w be congruent regular curves, so $w(t) = C(z(t))$ for some congruence C. Show that the evolutes z_*, w_* satisfy the relation $w_*(t) = C(z_*(t))$, so are likewise congruent.

8.2.13 Let z be a regular curve with speed s and curvature κ. Show that at parameters t where $\kappa'(t) \neq 0$ the curvature κ_* of the evolute is given by the formula

$$\kappa_*(t) = \frac{s\kappa^3(t)}{|\kappa'(t)|}.$$

8.2.14 Write down formulas for the evolute of a graph $y = f(x)$. Use your formulas to show that if the trace of the evolute is contained in the x-axis then the function f satisfies the differential equation $1 + ff'' + f'^2 = 0$. By integrating this equation, deduce that the trace of f is an arc of a circle.

8.3 Parallels

In the previous section the evolute of a curve was introduced as the locus of centres of curvature. There is however a more dynamic way of viewing the evolute, via the concept of 'parallel' curves. Let z be a regular curve, and let d be a real number. By the *parallel curve* z_d at distance d we mean the curve defined by

$$z_d(t) = z(t) + dN(t)$$

where $N(t)$ is the unit normal vector at t. Thus z_d is the locus of points distance d from z measured along the normal line. The components x_d, y_d are given by the following formulas, where x, y are the components of z, and we suppress the parameter t for economy of expression:

$$x_d = x - \frac{dy'}{\sqrt{x'^2 + y'^2}}, \quad y_d = y + \frac{dx'}{\sqrt{x'^2 + y'^2}}.$$

The terminology is based on the facts that the parallel curves to a line are the parallel lines in the sense of Example 1.13: likewise, the parallel curves to a circle are its concentric circles. (Exercise 8.3.1.) In

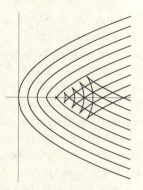

Figure 8.10. Parallels of a parabola

both these instances the parallel curves are similar to the given one. The next example illustrates the initially surprising fact that in general the parallels fail to be similar to the original curve.

Example 8.14 Consider the parabola $x(t) = at^2$, $y(t) = 2at$ with $a > 0$. The reader will readily check that the parallel at distance d is given by

$$x_d(t) = at^2 - \frac{d}{\sqrt{1+t^2}}, \quad y_d(t) = 2at + \frac{dt}{\sqrt{1+t^2}}.$$

The parallel z_d meets the x-axis when $y_d = 0$, i.e. when $t = 0$ or $\sqrt{1+t^2} = -d/2a$. In the former case the point of intersection is $(-d, 0)$. The latter relation is satisfied for some $t \neq 0$ if and only if $d < -2a$ in which case we obtain a second *distinct* point of intersection. (Figure 8.10.) For $d \geq -2a$ they do bear more than a passing resemblance to a parabola, but for $d < -2a$ they exhibit new features, namely a self crossing on the x-axis, and two irregular points, both of which can be shown to be ordinary cusps. (Exercise 8.3.2.)

The reader with access to a computer algebra plotting program can trace parallels of favourite curves on an experimental basis. Figure 8.11 is another illustration, namely some of the parallels of an ellipse.

Example 8.15 In dealing with parallels one must not forget that they are only defined for *regular* curves. Consider for instance the astroid $x(t) = \cos^3 t$, $y(t) = \sin^3 t$. The curve is periodic with period 2π, and in the interval $0 \leq t < 2\pi$ has four irregular parameters $t = 0$, $t = \pi/2$, $t = \pi$, $t = 3\pi/2$: deleting these four points we obtain four open intervals

8.3 Parallels

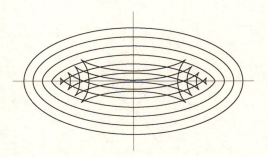

Figure 8.11. Parallels of an ellipse

on each of which the restriction is regular. Figure 8.12 illustrates for each of the four restrictions the two parallels at distances $d = 1/2$, $d = -1/2$, yielding eight-curves in all. The reader is urged to pause and clarify which curves represent parallels at distance $d = 1/2$, and which represent parallels at distance $d = -1/2$. The point of the example is that it is all too easy to misinterpret the illustration.

Comparing Figure 8.10 with Figure 8.2 suggests that the locus of irregular points for the parallels is the evolute. In fact that is the case, provided the statement is worded a little more carefully.

Lemma 8.6 *Let z be a regular curve, and let d be a real number. Then a parameter t is irregular for the parallel z_d if and only if $z_d(t) = z_*(t)$.*

Proof It is no restriction to assume that z has unit speed. Using the Serret–Frenet Formulas (5.4) we see that the tangent vector to the parallel z_d at t is

$$z'_d(t) = z'(t) + dN'(t) = z'(t) - d\kappa(t)z'(t) = (1 - d\kappa(t))z'(t) \qquad (8.5)$$

As $z'(t) \neq 0$ it follows that t is irregular for z_d if and only if $1 - d\kappa(t) = 0$, i.e. if and only if $z_d(t) = z_*(t)$. □

Thus the irregular parameters t for the parallel z_d are precisely those for which $z_d(t)$ is the corresponding point $z_*(t)$ on the evolute.

Example 8.16 It adds to one's understanding to view the evolute as a transition between two distinct types of behaviour. Provided $d\kappa(t) \neq 1$ the displayed relation (8.5) in the above proof shows that the tangent vectors $z'_d(t)$, $z'(t)$ are linearly dependent. Let T, T_d be the unit tangent vectors,

Figure 8.12. Two parallels of an astroid

and N, N_d be the unit normal vectors for z, z_d. Then for $d\kappa(t) < 1$ the tangent vectors are parallel in the same direction, and $T = T_d$, $N = N_d$: on the other hand, for $d\kappa(t) > 1$ the tangent vectors are parallel in opposite directions, and $T = -T_d$, $N = -N_d$. More succinctly, we can write $T_d = \epsilon T$, $N_d = \epsilon N$, where $\epsilon = +1$ when $d\kappa < 1$ and $\epsilon = -1$ when $d\kappa > 1$. In particular, if t is a regular parameter for both curves the normal lines at these points will coincide. Thus the normal lines for any parallel z_d coincide with the normal lines for the original curve z.

Lemma 8.7 *Let t be a parameter which is regular for both a curve z and a parallel z_d. Then the curvatures κ, κ_d of z, z_d are related by*

$$\kappa_d = \frac{\epsilon \kappa}{1 - d\kappa}.$$

Proof Let T, N, s be respectively the unit tangent vector, unit normal vector and speed for z: and let T_d, N_d, s_d be the corresponding quantities for the parallel z_d. By Lemma 8.6 we need only consider parameters for which $d\kappa \neq 1$. As in the above discussion we can write $T_d = \epsilon T$, $N_d = \epsilon N$. Differentiation then yields $T_d' = \epsilon T'$, $N_d' = \epsilon N'$. From the displayed relation (8.5) we deduce that $s_d = \epsilon(1 - d\kappa)s$. Using the formula for the curvature in Section 5.2 we now have

$$\kappa_d = \frac{T_d' \bullet N_d}{s_d} = \frac{(\epsilon T') \bullet (\epsilon N)}{s_d} = \frac{(T') \bullet (N)}{\epsilon(1 - d\kappa)s} = \frac{\epsilon \kappa}{1 - d\kappa}.$$

□

Example 8.17 Let z be a regular curve. It follows immediately from the relation between the curvatures of z and its parallel z_d that a parameter t is inflexional for z if and only if it is inflexional for *all* the parallels z_d. In particular if z has no inflexions then *none* of the parallels z_d have inflexions.

Our final Lemma marries the concepts of evolute and parallel: paraphrased, it says that all the parallels of a curve z have the same evolute as z.

Lemma 8.8 *Let z be a regular curve without inflexions, and let t be a parameter which is regular for a parallel z_d. Then we have $(z_d)_*(t) = z_*(t)$.*

Proof Write N, N_d for the unit normal vectors associated to the curves z, z_d. As in the proof of Lemma 8.7 we can write $N = \epsilon N_d$. Then,

$$\begin{aligned} (z_d)_* &= z_d + \frac{N_d}{\kappa_d} \\ &= (z + dN) + \frac{(1 - d\kappa)}{\kappa} N \\ &= z + \frac{N}{\kappa} = z_*. \end{aligned}$$
□

Exercises

8.3.1 Show that the parallels of a regularly parametrized line are its parallel lines, and that the parallels of a regularly parametrized circle are its concentric circles.

8.3.2 Let z_d be the parallel of the parabola $x(t) = at^2$, $y(t) = 2at$ at distance d. (Example 8.14.) Show that z_d is regular if and only if $d > -2a$. Further, show that for $d < -2a$ it has exactly two irregular points, both ordinary cusps. What happens when $d = 2a$?

8.3.3 Let z be a regular curve, and let z_1^*, z_2^* be regular involutes starting respectively at the parameters t_1, t_2. Show that z_1^*, z_2^* are parallel curves. (Lemma 8.8 then tells us that z_1^*, z_2^* have the same evolute, confirming the result of Lemma 8.5.)

9
Vertices

The central object in the previous chapter is the evolute of a parametrized curve, the locus of centres of the circles of curvature. Recall that the circle of curvature has *at least* three point contact with the curve. In this chapter we will pursue these ideas to study the exceptional points on a curve where the circle of curvature actually has *at least* four point contact. Our first result is that such exceptional points correspond to stationary values of the curvature, the 'vertices' of the curve, enabling us to determine them in explicit examples. One of their virtues is that they tend to appear as highly visible points on a tracing of the evolute, whereas they may be effectively invisible on a tracing of the original curve. That emphasizes the point that the evolute picks up very subtle geometric information about a curve: indeed two visually similar curves may have quite dissimilar evolutes. It is for that reason that evolutes provide sensitive methods for distinguishing one curve from another, a matter of practical importance in some physical disciplines.

9.1 The Concept of a Vertex

Before proceeding to formalities it might be profitable to look at an explicit example in some detail.

Example 9.1 Consider the parabola z with components $x(t) = at^2$, $y(t) = 2at$ where $a > 0$. In Example 8.2 we showed that the circle of curvature at the parameter $t = 0$ has exactly four point contact with the parabola. In this example we will show that $t = 0$ is the *only* parameter for which the circle of curvature has four point contact. The contact

9.1 The Concept of a Vertex

function at the parameter t_0 with the circle of curvature at t_0 is

$$\gamma(t) = |z(t) - z_*(t_0)|^2 - |z(t_0) - z_*(t_0)|^2$$

where z_* is the evolute. In Example 8.5 we showed that z_* is the semicubical parabola with components $x_*(t) = 2a + 3at^2$, $y_*(t) = -2at^3$. Substituting for z, z_* in the displayed formula, and then successively differentiating with respect to t, we get

$$\gamma'(t) = 4a^2(t^3 - 3t_0^2 t + 2t_0^3), \quad \gamma''(t) = 12a^2(t^2 - t_0^2), \quad \gamma'''(t) = 24a^2 t.$$

Setting $t = t_0$ in these relations we obtain

$$\gamma(t_0) = 0, \quad \gamma'(t_0) = 0, \quad \gamma''(t_0) = 0, \quad \gamma'''(t_0) = 24a^2 t_0.$$

It follows that t_0 is a zero of γ of multiplicity ≥ 3, jumping to 4 if and only if $t_0 = 0$. In other words the circle of curvature always has at least three point contact, jumping to four point contact precisely when $t_0 = 0$.

In this example the significance of the parameter $t_0 = 0$ is that it corresponds to the unique point where the trace of the parabola meets the axis of symmetry, namely the vertex of the parabola. It would be grossly inefficient to have to repeat calculations such as that in Example 9.1 to find the points on a regular curve z for which the circle of curvature has higher contact. Our next result will provide a practical criterion. Recall first that the condition for at least k point contact is that the first $(k-1)$ derivatives of the contact function γ introduced in Section 8.1 should vanish. Thus we need explicit expressions for the first few derivatives of γ. For the circle C with centre c and radius r the contact function with C at the parameter t_0 was defined to be the smooth function

$$\gamma(t) = |z(t) - c|^2 - |z(t_0) - c|^2.$$

In Section 8.1 it was shown (Lemma 8.1) that the condition for at least two point contact is that c should lie on the normal line to z at t_0, and (Lemma 8.2) that at a non-inflexional parameter t_0 the condition for at least three point contact is that c should be the centre of curvature at t_0 defined by $c = z_*(t_0)$ with z_* the evolute. Pursuing this line of thought leads to the following result.

Lemma 9.1 *Let z be a regular curve, and let t be a non-inflexional parameter. The condition for the circle of curvature at t to have contact of order ≥ 4 is that $\kappa'(t) = 0$. And the condition for the contact to be of order ≥ 5 is that $\kappa'(t) = 0$, $\kappa''(t) = 0$.*

Proof For calculations we may assume z has unit speed. And for notational efficiency we suppress the parameter t. Thus the unit tangent vector is $z' = T$, and the Serret–Frenet Formulas read $T' = \kappa N$, $N' = -\kappa T$. Using these relations the reader is left to verify that the first four derivatives of γ are as follows.

$$\begin{aligned}
\gamma' &= 2(z-c) \bullet T \\
\gamma'' &= 2(z-c) \bullet (\kappa N) + 2 \\
\gamma''' &= 2(z-c) \bullet (-\kappa^2 T + \kappa' N) \\
\gamma'''' &= 2(z-c) \bullet (-3\kappa\kappa' T + (\kappa'' - \kappa^3)N) - 2\kappa^2.
\end{aligned}$$

In view of the above comments the condition for the contact at t to be ≥ 4 is that $\gamma'''(t) = 0$ with $c = z + (N/\kappa)$. Substituting for c in the formula for γ''', and using the relations $T \bullet N = 0$, $N \bullet N = 0$, we see immediately that the required condition is $\kappa'(t) = 0$. Likewise, for t to be a higher vertex the condition is that in addition we have $\gamma''''(t) = 0$ with $c = z + (N/\kappa)$. Further calculation shows that the required condition is $\kappa''(t) = 0$. □

On this basis we are led to the following definition. By a *vertex* of a curve z we mean a regular parameter t for which the curvature κ has a stationary value, i.e. $\kappa'(t) = 0$. In particular, local extrema of the curvature are vertices: and for curves of constant curvature (line segments and arcs of circles) every parameter is a vertex. When t is inflexional Lemma 7.9 tells us that t is a vertex if and only if t is undulational. And when t is non-inflexional a vertex is a parameter for which the circle of curvature at t has at least four point contact with the curve at t: vertices with exactly four point contact are *ordinary vertices*, and those with at least five point contact are *higher vertices*. Since contact of z with circles is invariant under changes of parameter, the same is true for the vertex concepts just introduced.

Example 9.2 When a curve z has exactly three point contact with the circle of curvature at a non-inflexional parameter t it crosses the circle close to t: and that is the case for contact of any odd order. However, for contact of even order the curve will stay on one side of the circle of curvature close to t. For instance the circle of curvature of the parabola $x = at^2$, $y = 2at$ with $a > 0$ has four point contact with the curve at $t = 0$, and the reader can see in Figure 8.1 how, close to the vertex, the parabola stays outside the circle of curvature.

9.1 The Concept of a Vertex

Lemma 9.1 tells us that the vertices of a curve arise as solutions of the equation $\kappa'(t) = 0$, in particular as the local extrema of $\kappa(t)$. For some examples this may be a reasonable calculation to attempt by hand.

Example 9.3 Consider the parabola $x(t) = at^2$, $y(t) = 2at$ with $a > 0$. The curvature κ and its derivative κ' are given by the formulas

$$\kappa(t) = \frac{-1}{2a(1+t^2)^{3/2}}, \quad \kappa'(t) = \frac{3t}{2a(1+t^2)^{5/2}}.$$

The derivative vanishes if and only if $t = 0$. There is therefore just one vertex, corresponding to the unique point $(0,0)$ where the parabola meets its axis of symmetry. (Incidentally, that is consistent with the terminology of Example 3.1.) Moreover the vertex is ordinary, as a brief calculation verifies that $\kappa''(0) = 3/2a \neq 0$.

Example 9.4 Consider the ellipse $x(t) = a\cos t$, $y(t) = b\sin t$, where $0 < b < a$. We saw in Example 5.3 that the curvature is given by

$$\kappa(t) = \frac{ab}{(a^2 \sin^2 t + b^2 \cos^2 t)^{3/2}}.$$

The reader is left to verify that the derivative of the curvature is given by

$$\kappa'(t) = \frac{3ab(b^2 - a^2)\sin t \cos t}{(a^2 \sin^2 t + b^2 \cos^2 t)^{5/2}}.$$

This vanishes if and only if $\sin t = 0$ or $\cos t = 0$. The relation $\sin t = 0$ corresponds to the points $(\pm a, 0)$ at the ends of the minor axis where the curvature assumes its minimum value a/b^2, whilst the relation $\cos t = 0$ corresponds to the points $(0, \pm b)$ at the ends of the major axis where the curvature assumes its maximum value b/a^2. There are therefore four vertices on the ellipse, corresponding to the extremities of its axes, agreeing with the terminology introduced in Example 3.2. The reader will readily check that all four vertices are ordinary, since in each case κ'' is non-zero.

Familiarity with simple examples (such as the standard conics) may lull one into thinking that the human eye is capable of picking out vertices on the trace of a curve. For instance it is very tempting to imagine that stationary points on the graph of a function $f(x)$ will correspond to vertices. However, the concept of a vertex is more subtle than that. It is easily verified (Exercise 9.1.7) that t is a vertex for the

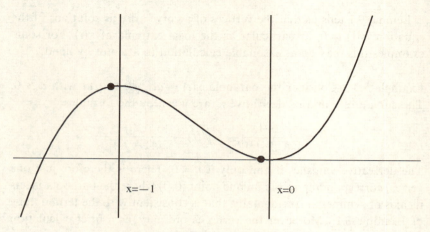

Figure 9.1. Vertices on the graph of $f(x) = x^2(2x+3)$

graph $z(t) = (t, f(t))$ if and only if the following condition is satisfied; in particular, a stationary parameter t is a vertex if and only if $f'''(t) = 0$.

$$f'''(t) = \frac{3f'(t)f''(t)^2}{1 + f'(t)^2}.$$

Example 9.5 Consider the graph of the function $f(x) = x^2(2x+3)$, parametrized as $x(t) = t$, $y(t) = t^2(2t+3)$. (Figure 9.1.) It is tempting to guess that the local extrema at $t = 0$, $t = -1$ are vertices. However, that is not the case, since identically $f'''(t) = 12 \neq 0$. A computer algebra program will verify that the relation displayed above holds for exactly two real values of t, approximately $t \approx -0.058$, $t \approx -1.058$. The corresponding points on the graph are marked on the figure by dots, neither of which springs to the eye as a candidate for a vertex.

As a prelude to the next example recall the statement of a basic calculus result, namely Rolle's Theorem. Let f be a smooth function defined on an open interval, and let a, b be real numbers with $a < b$ for which $f(a) = 0$, $f(b) = 0$; then there exists a t with $a < t < b$ for which $f'(t) = 0$. We wish to apply this to the curvature function κ associated to a regular curve z. In that situation Rolle's Theorem, combined with Lemma 9.1, says that *between any two inflexional parameters for z there is at least one vertex*. To illustrate these ideas, we consider the family of limacons whose inflexional behaviour was analysed in Examples 5.9 and 7.7, and summarized in Table 7.1.

9.1 The Concept of a Vertex

Example 9.6 We will determine how the number of vertices on the limacon $z(t) = 2e^{it} - he^{2it}$ depends on $h > 0$. The curvature and its derivative are given by

$$\kappa(t) = \frac{(2h^2 - 3h\cos t + 1)}{2(h^2 - 2h\cos t + 1)^{3/2}}, \quad \kappa'(t) = \frac{3h^2 \sin t\,(-h + \cos t)}{2(h^2 - 2h\cos t + 1)^{5/2}}.$$

Vertices appear when the derivative vanishes, i.e. when $\sin t = 0$ or $\cos t = h$. Now $\sin t$ vanishes only for $t = 0$ and $t = \pi$ in the range $0 \leq t < 2\pi$. Suppose first that $h \neq 1$. Then these parameters are regular and non-inflexional, hence vertices. The other possibility is $\cos t = h$. When $h > 1$ this has no solutions: and when $h < 1$ there are exactly two solutions in the range $0 \leq t < 2\pi$ (which are regular and non-inflexional) yielding two more vertices. In the exceptional case $h = 1$ of the cardioid only $t = \pi$ is a vertex, and $t = 0$ is the unique irregular parameter. Thus for $h < 1$ there are four vertices, for $h = 1$ there is one, and for $h > 1$ there are two. The reader is encouraged to correlate these conclusions with the inflexional behaviour described in Table 7.1, and the illustrations of limacons in Figure 3.9.

Example 9.7 In Exercise 5.4.5 it was shown that the eight-curve $x(t) = a\cos t$, $y(t) = a\sin t \cos t$ where $a > 0$ has just two inflexional parameters in the range $0 < t < 2\pi$, giving rise to inflexions at the unique self crossing. It is interesting to determine the number of vertices. A tedious calculation verifies that the derivative κ' of the curvature function vanishes if and only if

$$\sin t \cos^2 t\,(12\sin^4 t + 9\sin^2 t - 2) = 0.$$

We can discard the possibility $\cos t = 0$, as it gives rise to the two inflexions at the self crossing. When $\sin t = 0$ we obtain the two points $(\pm a, 0)$, other than the self crossing, where the curve meets the x-axis: these are therefore vertices. It remains to consider the expression in parentheses. To this end consider the function $\phi(S) = 12S^2 + 9S - 2$. Note that $\phi(-1) > 0$, $\phi(0) < 0$ and $\phi(1) > 0$ so in the interval $-1 \leq S \leq 1$ the equation $\phi(S) = 0$ has exactly one positive solution S^+, and exactly one negative solution S^-. Taking $S = \sin^2 t$, we see that the negative solution produces nothing, but the positive solution produces an equation $\sin^2 t = S^+$ with two roots for $\sin t$ giving rise to *four* values for t. That produces four more vertices, hence six in total.

Vertices

Exercises

9.1.1 Find the curvature functions associated to the parametrizations $x(t) = \pm a\cosh t$, $y(t) = b\sinh t$ of the branches of a standard hyperbola. Verify that neither branch has inflexions, and that each branch has exactly one vertex, corresponding to the intersection with the axis.

9.1.2 Determine the curvature and the vertices of the curve $x(t) = t^2$, $y(t) = t^5$.

9.1.3 Find the curvature of the curve $x(t) = 3t^2$, $y(t) = t - 3t^3$. Show that the curve has no inflexions, and precisely one vertex.

9.1.4 The general cycloid was introduced in Example 3.15 as the curve with components

$$x(t) = Rt - d\sin t, \quad y(t) = R - d\cos t.$$

Show that a parameter t is a vertex if and only if it satisfies one of the following relations, and use this fact to find all the vertices, for each value of the ratio d/R.

$$\sin t = 0, \quad \cos t = \frac{2d^2 - R^2}{dR}.$$

9.1.5 The piriform was introduced in Example 2.16 as the curve with the following components, where a, b are positive constants. Show that the piriform has a unique vertex at the parameter $t = 2a$.

$$x(t) = a(1 + \cos t), \quad y(t) = b\sin t(1 + \cos t).$$

9.1.6 Show that a regular parameter t of a curve z with components x, y is a vertex if and only if

$$(x'^2 + y'^2)(x'y''' - x'''y') = 3(x'x'' + y'y'')(x'y'' - x''y').$$

9.1.7 Use Exercise 9.1.6 to show that a parameter t is a vertex of a graph $z(t) = (t, f(t))$ if and only if

$$f'''(t) = \frac{3f'(t)f''(t)^2}{1 + f'(t)^2}.$$

Deduce that a parameter t with $f'(t) = 0$ is a vertex if and only $f'''(t) = 0$.

9.1.8 In each of the following cases use the result of Exercise 9.1.7 to show that the only vertices of the given graph occur at the

9.2 Appearance of Vertices on the Evolute

stated values of x.

(i) $y = \cosh x$, $\quad x = 0$
(ii) $y = \sinh x$, $\quad x = \log(\pm 1 + \sqrt{2})$
(iii) $y = e^x$, $\quad x = -\log \sqrt{2}$
(iv) $y = \log x$, $\quad x = 2^{-1/2}$
(v) $y = x^3$, $\quad x = 45^{-1/4}$
(vi) $y = x^4$, $\quad x = 0, \pm 56^{-1/6}$.

9.1.9 Use the result of Exercise 9.1.7 to show that the versiera given by the following formula (Example 3.7) has a unique vertex at $x = 0$:

$$f(x) = \frac{8a^3}{x^2 + 4a^2}.$$

9.1.10 Use the result of Exercise 9.1.7 to show that the Serpentine given by the following formula (Example 7.6) has exactly three vertices:

$$f(x) = \frac{b^2 x}{x^2 + a^2}.$$

9.1.11 Let z be a regular curve, and let t be an irregular parameter for the parallel z_d at distance d. Show that t is a cusp if and only if t is not a vertex of z. Further, show that in that case the cusp is ordinary if and only if t is not an inflexion for z.

9.2 Appearance of Vertices on the Evolute

For very simple curves (such as the standard parametrized conics) it is fairly clear to the eye which points arise from vertices. However, in general it is by no means easy to recognize vertices just by looking at the trace. It is therefore not without interest that vertices usually show up very clearly on the evolute. The main point to make in this section is that the evolute picks up very subtle geometric information from a curve. Here is the underlying theoretical result.

Lemma 9.2 *Let z be a regular curve, and let t_0 be a non-inflexional parameter. Then t_0 is a vertex for z if and only if t_0 is irregular for the evolute z_*. Moreover, if t_0 is an ordinary vertex for z then t_0 is an ordinary cusp for z_*.*

Proof We can assume that z has unit speed. It is no restriction to suppose that $t_0 = 0$, and since it is non-inflexional that $\kappa(0) \neq 0$. Recall that the evolute z_* is defined by $\kappa z_* = \kappa z + N$. (We suppress the parameter t.) Differentiating both sides of this relation, and using the Serret–Frenet Formulas, we obtain $\kappa^2 z'_* = -\kappa' N$. It follows immediately that $z'_*(0) = 0$ if and only if $\kappa'(0) = 0$. However, Lemma 9.1 tells us that $t = 0$ is a vertex for z if and only if $\kappa'(0) = 0$, so we deduce that $t = 0$ is irregular for the evolute z_* if and only if $t = 0$ is a vertex for z. That establishes the first statement in the result. Suppose now that $t = 0$ is an ordinary vertex of z, so by Lemma 9.1 we have $\kappa'(0) = 0$, $\kappa''(0) \neq 0$. Differentiating the above relation $\kappa^2 z'_* = -\kappa' N$ twice with respect to t, and setting $t = 0$, we get

$$\begin{cases} \kappa(0)^2 z''_*(0) &= -\kappa''(0) N(0) \\ \kappa(0)^2 z'''_*(0) &= 2\kappa(0)\kappa''(0) T(0) - \kappa'''(0) N(0). \end{cases}$$

The first relation tells us that $z''_*(0) \neq 0$, so $t = 0$ is a cusp. And the second shows that $z''_*(0)$, $z'''_*(0)$ are linearly independent, so $t = 0$ is an *ordinary* cusp. □

This result is well illustrated by the parabola (Example 9.3) and the ellipse (Example 9.4). In the case of the parabola there is a unique ordinary vertex giving rise to an ordinary cusp on the evolute (Figure 8.2) and in the case of the ellipse there are four ordinary vertices giving rise to four ordinary cusps on the evolute. (Figure 8.3.) Here is another example.

Example 9.8 Consider the cardioid $z(t) = 2e^{it} - e^{2it}$. In Example 7.11 we saw that the unique irregular parameter $t = 0$ is an ordinary cusp. And Example 9.6 tells us that $t = \pi$ is the unique vertex. By Example 8.10 the evolute z_* is a similar curve, related to the original by $3z_*(t) = -z(t+\pi)$.

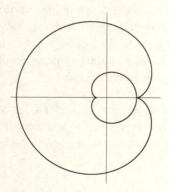

Figure 9.2. A cardioid and its evolute

Thus the evolute is another cardioid of one third the size, with $t = \pi$ the unique ordinary cusp arising from the vertex on z. Note incidentally that the evolute passes through the ordinary cusp on z, providing another illustration of a generality pointed out in Section 8.2, namely that the evolute of a curve necessarily 'passes through' its ordinary cusps. Both cardioids are illustrated in Figure 9.2.

9.3 The Four Vertex Theorem

In this chapter we have already applied basic calculus results to the curvature function to gain useful information about vertices. Here is another excursion into this line of thought. Any periodic curve z of period p can be viewed as the restriction to a compact interval of length p, and the same is true of the associated curvature function. One of the basic facts about such functions is that they have at least one local maximum, and at least one local minimum, and that such points are stationary. In particular the curvature function for a periodic curve has at least two stationary points, hence *a periodic curve has at least two vertices*. A good example of this is provided by limacons with constant $h > 1$ having *two* vertices; by contrast, limacons with constant $h < 1$ have *four* vertices. It is not chance that the fundamental difference between these cases lies in the fact that the former type have self crossings, whereas the latter do not. The key object in this section is an *oval*, by which we mean a regular periodic curve z having no self crossings. In the course of this book we have met several examples of ovals, such as circles, ellipses, limacons with $h < 1$, and some of the Lissajous figures.

We will impose a further condition on our curves, which (though mathematically unnecessary) has the merit of being geometrically compelling, and of leading to satisfyingly simple proofs. A regular curve z is said to be *convex* when the trace of the curve lies wholly on one side of the tangent line at any parameter t_0: more precisely, that means that given any parameter t_0 either $(z(t) - z(t_0)) \bullet N(t) \geq 0$ for all parameters t, or $(z(t) - z(t_0)) \bullet N(t) \leq 0$ for all parameters t. (Figure 9.3.) It is worth noting that a convex curve cannot have any ordinary inflexions, since at such a parameter the curve crosses the tangent line. (Lemma 7.8.)

Lemma 9.3 *Let z be a convex oval with period ω, and let L be a line meeting the trace at two distinct points $p_1 = z(t_1)$, $p_2 = z(t_2)$. Then either the whole line segment joining p_1, p_2 is contained in the trace of z, or z has no further intersections with L.*

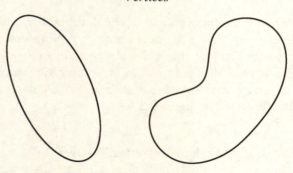

Figure 9.3. Convex and non-convex ovals

Proof We need a small preliminary observation, namely that if $p = z(\alpha)$, $q = z(\beta)$ are distinct points on the trace of a curve z, and L is a line with p on one side and q on the other, then L intersects the trace in at least one point $z(t)$. Let $x(t)$, $y(t)$ be the components of $z(t)$, let L have equation $ax + by + c = 0$, and set $f(t) = ax(t) + by(t) + c$. Then f is a continuous function for which $f(\alpha)$, $f(\beta)$ have different signs. It follows from the Intermediate Value Theorem that there exists at least one real number t in the interval $\alpha < t < \beta$ for which $f(t) = 0$, i.e. for which $z(t)$ lies on the line L.

Now we can proceed with the proof. When the whole line segment is contained in the trace there is nothing to prove, so assume that is not the case. Let $p_3 = z(t_3)$ be a further intersection of the trace with L. Note first that the tangent line at t_3 must coincide with L: otherwise p_1, p_2 lie on different sides of L, contradicting convexity. Next, choose any point $q \ne p_1, p_2$ on the line segment which does *not* lie in the trace. Consider any line M through q with p_1 on one side, and p_2 on the other. By the preliminary observation, the arc of z obtained by restriction to the interval $t_1 \le t \le t_2$ meets M at least once, at some point r: likewise, the arc of z obtained by restriction to the interval $t_2 \le t \le t_1 + \omega$ meets M at least once, at some point s. The points r, s are distinct, since z has no self crossings. And by convexity they must lie on the same side of the tangent line L. We can assume r is closer to q than s is. (Figure 9.4.) It remains to observe that by convexity all three points p_1, p_2, s must lie on the same side of the tangent to z at r: however, that is impossible, since r lies in the interior of the triangle formed by these three points. □

That brings us to the final result of this section, the Four Vertex Theorem of Mukhopadhaya.

9.3 The Four Vertex Theorem

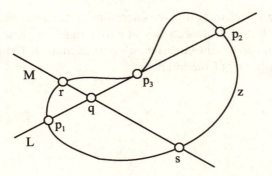

Figure 9.4. Proof of Lemma 9.3

Theorem 9.4 *Any convex oval z has at least four vertices.* (The Four Vertex Theorem.)

Proof Write $x(t)$, $y(t)$ for the components of $z(t)$, and ω for the common period of these mappings and their derivatives. We can assume z has unit speed. Further, we can assume the curvature κ is not constant: otherwise κ' is identically zero, and *every* parameter is a vertex. Since κ is continuous and periodic it has a global minimum at some parameter t_1, and a global maximum at another parameter $t_2 > t_1$. By calculus t_1, t_2 are stationary values of κ, hence vertices. It is no restriction to suppose that $t_1 = 0$, and by applying a congruence we can assume $y(t_1) = 0$, $y(t_2) = 0$. The segment of the x-axis joining $z(t_1), z(t_2)$ cannot be wholly contained in the trace of z: otherwise, κ is identically zero on that segment, and hence identically zero on the curve. It follows from Lemma 9.3 that t_1, t_2 are the only parameters t for which $y(t) = 0$, so that $y(t)$ has constant sign on the interval I_1 defined by $0 < t < t_2$, and the opposite sign on the interval I_2 defined by $t_2 < t < \omega$. The essence of the proof is to understand the sign changes of $\kappa'(t)$. The key observation is an integral equality based on the identity $x''(t) = -\kappa(t)y'(t)$, an immediate consequence of the Serret–Frenet equations for a unit speed curve. Using integration by parts for the last equality, we then have

$$0 = \int_0^\omega x''(t)dt = -\int_0^\omega \kappa(t)y'(t)dt = \int_0^\omega \kappa'(t)y(t)dt.$$

That can only hold when $\kappa'(t)y(t)$ changes sign. We claim that $\kappa'(t)$ cannot be of constant sign on I_1, and of constant sign on I_2. Indeed since $\kappa'(t)$ must assume positive values in I_1 (for the curvature to increase) and negative values in I_2 (for the curvature to decrease) the signs would be *opposite*. But $y(t)$ likewise has opposite signs on these intervals, so $\kappa'(t)y(t)$ would have constant sign, a contradiction. It follows that $\kappa'(t)$ changes sign in one of the intervals.

□

10
Envelopes

So far in this book we have been concerned solely with the geometry of a single parametrized curve z. But numerous situations give rise naturally to a *family* of curves (z_λ), where the parameter λ ranges over some set Λ. In this chapter we restrict ourselves to one parameter families of curves (meaning that Λ is an open interval of the real line) and study one feature of such a family, namely that it may have an 'envelope', roughly speaking a curve which at every point touches some curve in the family. A paradigm for the concept is the family of tangent lines to a given parametrized curve: in more detail, we start with a regular curve z, and for each parameter λ let z_λ be the tangent line to z at λ: then z has the property that at every parameter λ there is a curve in the family (z_λ) which touches it at that point. Our concern is with the reverse process, where one starts with a family of parametrized curves (z_λ), and seeks a parametrized curve z with the property that at every point it touches some curve in the family. The key result of this chapter is the Envelope Theorem, which in principle enables one to find all possible envelopes of a given family of curves.

10.1 Envelopes

To make sense of the idea of an 'envelope' we require some formal definitions. By a *one parameter family* Z of parametrized curves we mean a smooth mapping $Z : \Lambda \times I \to \mathbb{R}^2$, where Λ, I are open intervals of the real line. It is helpful to think of $\Lambda \times I$ as a subset of the (λ, t)-plane. The *members* of the family are the parametrized curves $z_\lambda : I \to \mathbb{R}^2$ defined by the formula $z_\lambda(t) = Z(\lambda, t)$. Thus we think of the 'vertical' lines in the (λ, t)-plane parametrized as $t \mapsto (\lambda, t)$, and z_λ as the result of composing Z with this parametrized curve. (Figure 10.1.)

138   Envelopes

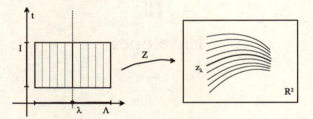

Figure 10.1. A family of curves

Example 10.1 Consider the family of circles centred on the line $y = x$ and touching both the x-axis and the y-axis. (Figure 10.2.) Take a point (λ, λ) on the line $y = x$. The circle centre (λ, λ) and radius $|\lambda|$ can be parametrized as

$$z_\lambda(t) = Z(\lambda, t) = (\lambda + \lambda \cos t, \lambda + \lambda \sin t)$$

defining a family $Z : \mathbb{R} \times \mathbb{R} \to \mathbb{R}^2$. Note that when $\lambda = 0$ the trace of z_λ is the origin, representing a degenerate member of the family.

Let $Z : \Lambda \times I \to \mathbb{R}^2$ be a family of parametrized curves, and let $e : U \to \Lambda \times I$ be a regular parametrized curve with domain U an open interval. Write $e(u) = (\lambda(u), t(u))$, and $E(u) = Z(\lambda(u), t(u))$ for the composite with Z. We call E an *envelope* (and e a *pre-envelope*) for the family Z, when the following conditions are satisfied.

(i) The function λ is non-constant on any non-trivial subinterval of U. (*The Variability Condition*.)

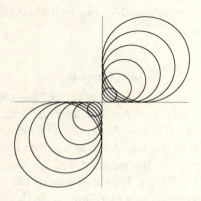

Figure 10.2. A family of circles

10.1 Envelopes

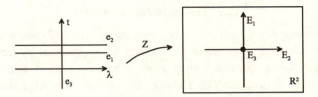

Figure 10.3. Two envelopes of the family of circles

(ii) For all u the curve E is tangent at u to the curve $z_{\lambda(u)}$ at the parameter $t(u)$, meaning that the tangent vectors $E'(u)$, $z'_{\lambda(u)}(t(u))$ are linearly dependent. (*The Tangency Condition*.)

Example 10.2 Let us return to the family of circles Z in Example 10.1 defined by

$$z_\lambda(t) = (\lambda + \lambda \cos t, \lambda + \lambda \sin t).$$

Consider the regular curves $e_1(u) = (u, \pi)$, $e_2(u) = (u, 3\pi/2)$, $e_3(u) = (0, u)$ in $\mathbb{R} \times \mathbb{R}$. (Figure 10.3.) The traces of e_1, e_2 are the 'horizontal' lines in $\mathbb{R} \times \mathbb{R}$ defined by $t = \pi$, $t = 3\pi/2$ and the trace of e_3 is the 'vertical' line in $\mathbb{R} \times \mathbb{R}$ defined by $\lambda = 0$. The reader is left to check that the composites of e_1, e_2, e_3 with Z are the curves $E_1(u) = (0, u)$, $E_2(u) = (u, 0)$, $E_3(u) = (0, 0)$. The trace of E_1 is the y-axis, the trace of E_2 is the x-axis, and the trace of E_3 is the origin. Write $\lambda_1, \lambda_2, \lambda_3$ for the first components of e_1, e_2, e_3, and t_1, t_2, t_3 for the second components. Of the functions $\lambda_1(u) = u$, $\lambda_2(u) = u$, $\lambda_3(u) = 0$ only the last fails to satisfy the Variability Condition, so E_3 fails to be an envelope. It remains to consider the tangency condition. Note first that $z'_\lambda(t) = (-\lambda \sin t, \lambda \cos t)$. At any parameter u the tangent vectors $E'_1(u) = (0, 1)$, $z'_\lambda(\pi) = (0, -\lambda)$ are linearly dependent, so e_1 is a preenvelope, and E_1 is an envelope. Also, the tangent vectors $E'_2(u) = (1, 0)$, $z'_\lambda(3\pi/2) = (\lambda, 0)$ are linearly dependent, so likewise e_2 is a pre-envelope, and E_2 is an envelope.

Example 10.3 Consider the family of parametrized curves defined by $Z(\lambda, t) = (t, \lambda(t - \lambda)^2)$. For $\lambda = 0$ the curve z_λ parametrizes the x-axis, whilst for $\lambda \neq 0$ it parametrizes a parabola, tangent to the x-axis at the point $(\lambda, 0)$. For the regular curve $e(u) = (u, u)$ we have $E(u) = (u, 0)$, with trace the x-axis. Evidently E is an envelope for Z. This example makes the important point that *a member of the family* (z_λ) *is not precluded from being an envelope of the family*.

10.2 The Envelope Theorem

So far we have simply 'spotted' envelopes using no more than a modicum of native wit. The next stage is to describe a practical procedure for finding envelopes, provided they exist. The key idea is embodied in the following definition. The *singular set* of a family of parametrized curves $Z : \Lambda \times I \to \mathbb{R}^2$ is the subset of the domain $\Lambda \times I$ defined by the equation $\det J(Z) = 0$, where $J(Z)$ is the *Jacobian matrix* of the mapping $Z(\lambda, t) = (X(\lambda, t), Y(\lambda, t))$ defined as follows, where X_λ, Y_λ are the derivatives of X, Y with respect to λ, and X_t, Y_t the derivatives with respect to t.

$$J(Z) = \begin{pmatrix} X_\lambda & Y_\lambda \\ X_t & Y_t \end{pmatrix}.$$

Put another way, the singular set is the set of points (λ, t) for which the vectors Z_λ, Z_t are defined and linearly dependent. When we use complex notation that is the same thing as saying that the quotient of Z_λ, Z_t is a *real* number. In practice this can be a more convenient way of phrasing the determinantal condition.

Example 10.4 For the family $Z : \mathbb{R} \times \mathbb{R} \to \mathbb{R}^2$ of Example 10.1 we have $X(\lambda, t) = \lambda + \lambda \cos t$, $Y(\lambda, t) = \lambda + \lambda \sin t$ and the Jacobian determinant is given by $\det J(Z) = -\lambda(1 + \sin t + \cos t)$. This vanishes when $\lambda = 0$ or $1 + \sin t + \cos t = 0$. The former relation defines the t-axis in the (λ, t)-plane. The latter relation is satisfied either when $\sin t = 0$ and $\cos t = -1$ or when $\sin t = -1$ and $\cos t = 0$: in the former case $t = (2n-1)\pi$, and in the latter case $t = (2n - \frac{1}{2})\pi$ where n is any integer. Thus the singular set is the union of the t-axis with two infinite families of 'horizontal' lines in the (λ, t)-plane.

Theorem 10.1 *Let $Z : \Lambda \times I \to \mathbb{R}^2$ be a family of parametrized curves, let U be an open interval, and let $e : U \to \Lambda \times I$ be a regular curve satisfying the Variability Condition. Then e is a pre-envelope for Z (and E is an envelope) if and only if the trace of e lies in the singular set of Z.* (The Envelope Theorem.)

Proof Differentiating the formulas for E, z_λ we obtain the relations

$$\begin{aligned} E'(u) &= \lambda'(u) Z_\lambda(\lambda(u), t(u)) + t'(u) Z_t(\lambda(u), t(u)) \\ z'_{\lambda(u)}(t(u)) &= Z_t(\lambda(u), t(u)). \end{aligned}$$

10.2 The Envelope Theorem

Suppose first that $e(u) = (\lambda(u), t(u))$ lies in the singular set of Z, so the vectors Z_λ, Z_t are linearly dependent at $(\lambda(u), t(u))$. It follows immediately from the displayed relations that $E'(u)$, $z'_{\lambda(u)}(t(u))$ are linearly dependent, hence that E is tangent at u to the curve $z_{\lambda(u)}$ at $t(u)$. Conversely, suppose that E is tangent at u to the curve $z_{\lambda(u)}$ at $t(u)$, so the vectors $E'(u)$, $z'_{\lambda(u)}(t(u))$ are linearly dependent. Provided $\lambda'(u) \neq 0$ it follows immediately from the displayed formulas that the vectors Z_λ, Z_t are linearly dependent at $(\lambda(u), t(u))$, hence that $e(u) = (\lambda(u), t(u))$ lies in the singular set of Z. It remains to establish the same conclusion at points $(\lambda(u), t(u))$ where $\lambda'(u) = 0$. It is at this juncture that we use the Variability Condition. Recall from calculus that a smooth function is constant on an open interval if and only if its derivative is identically zero on that interval. Since λ is assumed to be non-constant on any non-trivial subinterval of U, that means that its derivative λ' cannot be identically zero on any non-trivial subinterval. Thus given a parameter u for which $\lambda'(u) = 0$ there must be parameters v, with v arbitrarily close to u, for which $\lambda'(v) \neq 0$, and hence for which $e(v) = (\lambda(v), t(v))$ lies in the singular set of Z. It remains only to observe that the singular set of Z is closed, so $(\lambda(u), t(u))$ must also lie in it. $\square$

The Envelope Theorem provides a computational device for determining the envelopes of a given family of curves $Z(\lambda, t)$. The method is first to determine the singular set by analysing the equation $\det J(Z) = 0$. In principle it will comprise a number of 'curves'. We discard all 'vertical' line segments which appear in the singular set, and then parametrize the remaining curves to obtain pre-envelopes e, each giving rise to an envelope E.

Example 10.5 Let us return once again to Example 10.1 in the light of the Envelope Theorem. By Example 10.4 the singular set of the family is the union of the t-axis, and two infinite families of 'horizontal' lines $t = (2n-1)\pi$ and $t = (2n-\frac{1}{2})\pi$ where n is any integer. The t-axis can be discarded as any regular parametrization fails to satisfy the Variability Condition. Under the mapping Z each line $t = (2n-1)\pi$ is mapped to the y-axis, and each line $t = (2n-\frac{1}{2})\pi$ is mapped to the x-axis. Thus we recover the two envelopes of Example 10.2.

In the previous example we did not find any envelopes other than the 'obvious' ones described in Example 10.2, nor did we expect to. However, one has to be careful. The point of the next example is that we have a

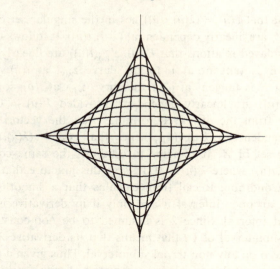

Figure 10.4. Astroid as an envelope of ellipses

family of curves with one 'obvious' envelope where our methods produce a second envelope which is far from being 'obvious'.

Example 10.6 Consider again the family $Z(\lambda, t) = (t, \lambda(t-\lambda)^2)$ of parabolas discussed in Example 10.3. The reader will readily check that

$$\det J(Z) = -(t - \lambda)(t - 3\lambda)$$

so the singular set of Z is the union of the lines $t = \lambda$ and $t = 3\lambda$. The line $t = \lambda$ gives rise to the 'obvious' envelope of Example 10.3, namely the x-axis. However by Theorem 10.1 the line $t = 3\lambda$, parametrized as $e(u) = (u, 3u)$, gives rise to a *second* envelope $E(u) = (3u, 4u^3)$, a parametrization of the cubic curve $27y = 4x^3$.

Example 10.7 Consider the family of ellipses defined by $x_\lambda(t) = \lambda \cos t$, $y_\lambda(t) = (1 - \lambda) \sin t$. (Of course for $\lambda = 0, 1$ these ellipses degenerate to closed line segments.) The reader will readily check that the singular set of the family is given by $\lambda = \cos^2 t$. Substituting in the formulas for the family we see that the envelope is $x(t) = \cos^3 t$, $y(t) = \sin^3 t$ which is an astroid. (Figure 10.4.)

Example 10.8 Consider the family of circles whose centre lies on the unit circle, and which have the property that they are tangent to the x-axis.

10.2 The Envelope Theorem

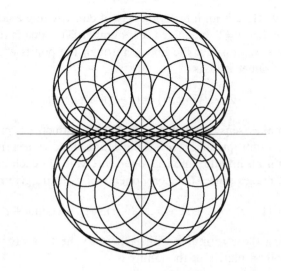

Figure 10.5. Nephroid as an envelope of circles

We parametrize the unit circle as $z(\lambda) = e^{i\lambda}$, so the circle centred at $z(\lambda)$ and tangent to the x-axis can be parametrized as

$$Z(\lambda, t) = z_\lambda(t) = e^{i\lambda} + (\sin \lambda)e^{it}.$$

Our objective is to show that the envelope of this family of circles is a nephroid. (Figure 10.5.) The components of $z_\lambda(t)$ are given by

$$x_\lambda(t) = \cos \lambda + \sin \lambda \cos t, \qquad y_\lambda(t) = \sin \lambda + \sin \lambda \sin t.$$

Differentiating these formulas with respect to the two variables we see that the Jacobian matrix of the family is

$$J(Z) = \begin{pmatrix} \cos \lambda \cos t - \sin \lambda & -\sin \lambda \sin t \\ \cos \lambda \sin t + \cos \lambda & \sin \lambda \cos t \end{pmatrix}.$$

The reader will easily verify that the determinant vanishes if and only if $\sin \lambda = 0$ or $\sin(\lambda - t) = \cos \lambda$. In the (λ, t)-plane the set defined by the first relation $\sin \lambda = 0$ comprises infinitely many 'vertical' lines ($\lambda = 2n\pi$ with n an integer) which do not contribute to the envelope. The second relation leads to $\cos t = \sin 2\lambda$, $\sin t = -\cos 2\lambda$ and hence $e^{it} = -ie^{2i\lambda}$. Substituting for e^{it} in the formula for the family we see that the envelope is parametrized as a scalar multiple of the 'standard' nephroid $3e^{i\lambda} - e^{3i\lambda}$.

Example 10.9 Here is an interesting way of constructing epicycloids and hypocycloids. Let $n \ne 1$ be a fixed real number. The idea is that for each real number λ we consider the line z_λ through the points $e^{i\lambda}$, $e^{in\lambda}$ on the unit circle, parametrized as

$$Z(\lambda,t) = z_\lambda(t) = (1-t)e^{i\lambda} + te^{in\lambda}.$$

That only makes sense provided $e^{i\lambda}$, $e^{in\lambda}$ are distinct, i.e. provided λ is not an integer multiple of $2\pi/(n-1)$: deleting these values from the real line we obtain an infinite union of open intervals, on each of which the construction makes sense. The components of $z_\lambda(t)$ are given by

$$x_\lambda(t) = (1-t)\cos\lambda + t\cos n\lambda, \quad y_\lambda(t) = (1-t)\sin\lambda + t\sin n\lambda.$$

Differentiating these formulas with respect to the two variables we see that the Jacobian matrix of the family is

$$J(Z) = \begin{pmatrix} -(1-t)\sin\lambda - tn\sin n\lambda & (1-t)\cos\lambda + tn\cos n\lambda \\ \cos n\lambda - \cos\lambda & \sin n\lambda - \sin\lambda \end{pmatrix}.$$

The reader will readily verify that the determinant of the Jacobian matrix for the family is given by

$$\det J(Z) = (nt + t - 1)\{\cos\lambda(n-1)) - 1\}.$$

The second factor only vanishes when λ takes a constant value (so does not give rise to an envelope) whilst the first factor vanishes if and only if $(n+1)t = 1$. When $n = -1$ there is no solution, and hence no envelope: geometrically that is not too surprising since then all the lines z_λ are parallel to the 'vertical' axis. Provided $n \ne -1$ the first factor vanishes on the parametrized line $\lambda = u$, $t = 1/(n+1)$, and gives rise to an envelope

$$E(u) = \frac{1}{n+1}\left\{ne^{iu} + e^{inu}\right\}.$$

When $n = 0$ all the lines z_λ pass through the point $(1,0)$, and the envelope is a constant curve with trace that point. However when $n \ne 0$ we see from Example 6.9 that (modulo a scalar factor) the envelope is equivalent to a trochoid (3.2) with $h = 1$ and ratio $n - 1$: indeed for $n > 1$ we have an epicycloid, and for $n < 1$ a hypocycloid. It follows that we can obtain *any* epicycloid or hypocycloid via this construction. Figure 10.6 illustrates the case $n = 2$ when the envelope is a cardioid.

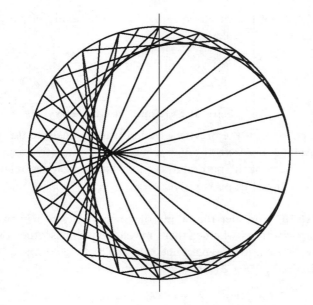

Figure 10.6. Cardioid as an envelope of lines

Exercises

10.2.1 Determine the envelopes of the family of circles of radius 1 centre on the x-axis.

10.2.2 Show that the envelopes of the family of circles $x(t) = \lambda + a\cos t$, $y(t) = -\lambda + a\sin t$ with $a > 0$ are the lines $x + y \pm \sqrt{2a} = 0$.

10.2.3 In each of the following cases determine the envelope of the given family of parametrized curves:

(i) $x(t) = t,$ $y(t) = \lambda - \lambda^2 t$
(ii) $x(t) = t,$ $y(t) = \lambda^2 - \lambda t$
(iii) $x(t) = t,$ $y(t) = \lambda^2 - \lambda^2 t$
(iv) $x(t) = t,$ $y(t) = 27\lambda - \lambda^3 t^2$
(v) $x(t) = t,$ $y(t) = \lambda^3 + \lambda t.$

10.2.4 In each of the following cases determine the envelope of the given family of parametrized curves:

(i) $x(t) = t,$ $y(t) = t^2 + 2\lambda t + \lambda + \lambda^2$
(ii) $x(t) = \lambda + t,$ $y(t) = \lambda t^2$
(iii) $x(t) = \lambda + t^2,$ $y(t) = \lambda t$
(iv) $x(t) = t^2 + \lambda,$ $y(t) = \lambda t.$

10.2.5 Show that the envelope of the following family of parametrized curves is a deltoid.

$$Z(\lambda, t) = (t^2 - \lambda^2 + 2t, 2t\lambda - 2\lambda).$$

10.3 Natural Envelopes in Geometry

The main purpose of the previous section was to illustrate the workings of the Envelope Theorem through a number of attractive examples. In this section we look at natural ways in which envelopes relate to the geometry developed in previous chapters.

Example 10.10 Consider the family of tangent lines to a regular curve z. The question arises whether there may be envelopes other than z for this family. Write λ for the parameter on z. The tangent line at λ can be parametrized as $z_\lambda(t) = Z(\lambda, t) = z(\lambda) + tz'(\lambda)$. Here

$$Z_\lambda(\lambda, t) = z'(\lambda) + tz''(\lambda), \quad Z_t(\lambda, t) = z'(\lambda).$$

Recall that the singular set is defined by the condition that the quotient of Z_λ, Z_t is real, equivalently that $tz''(\lambda)/z'(\lambda)$ is real, i.e. if and only if $t = 0$ or $z'(\lambda)$, $z''(\lambda)$ are linearly dependent, i.e. if and only if $t = 0$ or $\kappa(\lambda) = 0$, where κ denotes the curvature. Thus in the (λ, t)-plane the singular set of Z comprises the union of the λ-axis and 'vertical' lines $\lambda = \lambda_0$, one for each inflexional parameter λ_0. We can discard the 'vertical' lines as they fail to satify the Variability Condition. By the Envelope Theorem the λ-axis, parametrized as $e(u) = (u, 0)$, gives rise to the envelope $E(u) = z(u)$, which is indeed the original curve.

The point of the next example is that it shows that parallel curves can be usefully described as envelopes in a natural way.

Example 10.11 Consider the family of circles of fixed radius $d > 0$ centred at points on a regular curve z. Write λ for the parameter on z. The circle of radius d centred at $z(\lambda)$ can be parametrized as $z_\lambda(t) = Z(\lambda, t) = z(\lambda) + de^{it}$. Here

$$Z_\lambda(\lambda, t) = z'(\lambda), \quad Z_t(\lambda, t) = die^{it}.$$

The quotient of the derivatives is real if and only if e^{it} is orthogonal to the tangent vector $z'(\lambda)$, i.e. if and only if $e^{it} = \pm N(\lambda)$, with $N(\lambda)$ the normal vector to z at the parameter λ. Substituting for e^{it} in Z we obtain

10.3 Natural Envelopes in Geometry

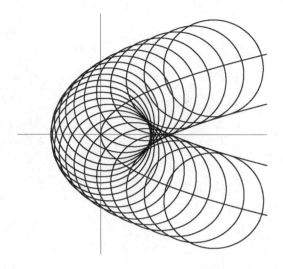

Figure 10.7. Parabola parallel as a circle envelope

two envelopes $z(\lambda) \pm dN(\lambda)$, which we recognize as the two parallels of z at distance d. (Section 8.3.)

The value of this example is that given a physical tracing of a curve z on a sheet of paper, it provides a practical method for tracing the curves parallel to z at given distance d. Using compasses to trace a large number of circles of radius d centred on the curve, the parallels become visible. (It is of course more efficient to trace the curve and the circles via a computer algebra program and display the result on a computer screen.) A good example is provided by a parabola, where the parallels exhibit interesting and unexpected features. (Example 8.14 and Figure 10.7.)

Finally, we make the key connexion between the concepts of evolute and envelope, which will be exploited in Chapter 12 when we discuss caustics of plane curves.

Lemma 10.2 *Let z be a regular curve z having no inflexions. Then the evolute of z is an envelope for the family of normal lines.*

Proof Write λ for the parameter on the curve. The normal line at λ is parametrized as

$$z_\lambda(t) = Z(\lambda, t) = z(\lambda) + tN(\lambda)$$

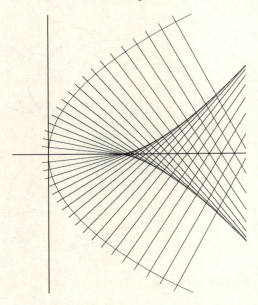

Figure 10.8. Envelope of the normals for a parabola

with $N(\lambda) = iz/s$ the normal vector at λ, where s is the speed of z. For this example the components X, Y of Z are given by

$$X(\lambda, t) = x(\lambda) - \left(\frac{t}{s}\right) y(\lambda), \quad Y(\lambda, t) = y(\lambda) + \left(\frac{t}{s}\right) x(\lambda)$$

and calculation yields $\det J(Z) = s\{1 - t\kappa(\lambda)\}$, with $\kappa(\lambda)$ the curvature. The Jacobian determinant vanishes on the set defined by $1 = t\kappa(\lambda)$. Since z has no inflexions the curvature is nowhere zero, so we can rewrite this as $t = 1/\kappa(\lambda)$ with the regular parametrization $e(u) = (\lambda(u), t(u))$ where $\lambda(u) = u$, $t(u) = 1/\kappa(u)$. By Theorem 10.1 the curve e is a pre-envelope for the family. The corresponding envelope is defined by the following relation, which according to (8.3) is the evolute of z.

$$E(u) = z(u) + \frac{N(u)}{\kappa(u)}.$$

□

Example 10.12 An illustration of this example is provided by the parabola $x(t) = at^2$, $y(t) = 2at$ with $a > 0$. According to Example 8.5 the evolute is the semicubical parabola $x_*(t) = 2a + 3at^2$, $y_*(t) = -2at^3$ illustrated in Figure 10.8 as the envelope of the family of normals.

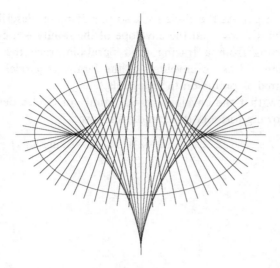

Figure 10.9. Envelope of the normals for an ellipse

Example 10.13 Another good example is provided by the ellipse $x(t) = a\cos t$, $y(t) = b\sin t$ with $0 < b < a$. According to Example 8.6 the evolute is the following curve, obtained from an astroid by scaling just one of the variables. The evolute is illustrated in Figure 10.9 as the envelope of the family of normals.

$$x_*(t) = \left(\frac{a^2-b^2}{a}\right)\cos^3 t, \qquad y_*(t) = \left(\frac{b^2-a^2}{b}\right)\sin^3 t.$$

Exercises

10.3.1 Let L be a line distinct from the axes, and let X, Y be respectively the points where L intersects the positive x-, y- axes. Determine the envelope of the family of such lines for which the triangle OXY has constant area $2k^2$.

10.3.2 Let L be a line distinct from the axes, and let X, Y be respectively the points where L intersects the positive x-, y- axes. Determine the envelope of the family of such lines for which the distance XY has constant length k.

10.3.3 A circle of radius 1 rolls along the line $y = -1$. Consider that diameter of the circle which is the x-axis when the centre is at the origin. As the circle rolls, so this diameter describes a family of lines. Show that the envelope of the family is a cycloid.

10.3.4 Starting from a tracing of a parabola trace the parallels at distance d as an envelope of the family of circles of radius d centred at points on the parabola.

10.3.5 Show that the envelope of the family of circles defined by the following formula is a standard ellipse:

$$Z(\lambda, t) = (\cos \lambda + \sin \lambda \cos t, \sin \lambda \sin t).$$

11
Orthotomics

In this chapter we describe the 'orthotomic' curves associated to a regular curve z, and a choice of point q, known as the *source*. The motivation lies in the material of Chapter 12, where we study 'caustics' which arise when a beam of light emanating from a point source q is reflected from a planar 'mirror', represented by the curve z. The key result will be that the caustic is the evolute of the orthotomic. However, orthotomics are of interest in their own right, and merit separate discussion. In Section 11.2 we present a formula for orthotomics, and use this to exemplify the surprisingly complex geometry which arises from the orthotomics of some quite simple curves. And in Section 11.5 we show how to use the envelope construction to reverse the process of constructing orthotomics.

11.1 Reflexions

Our starting point is to extend the discussion of isometries begun in Chapter 6. Here is the underlying mental picture for a reflexion. Suppose we are given a line L, and a point p. How do we construct a point q such that L is the orthogonal bisector of the line segment joining p, q?

Lemma 11.1 *Let L be a line and let p be a point. There is a unique point $q = R(p)$ with the property that every point z on L is equidistant from p, q. Moreover, q is given by the following formulas, where u is a point on L, T is a unit vector in the direction of L, and N is a unit vector orthogonal to the direction of L.*

$$R(p) = (2u - p) + 2\{(p - u) \bullet T\}T = p - 2\{(p - u) \bullet N\}N. \qquad (11.1)$$

Proof Let w be the orthogonal projection of p onto L. Figure 1.3 suggests that the point q defined by $q = 2w - p$ has the required properties: indeed, since (Lemma 1.3) the vectors $(w - p)$, $(w - z)$ are orthogonal

$$\begin{aligned} |z - q|^2 &= |z - (2w - p)|^2 = |(z - p) - 2(w - p)|^2 \\ &= |z - p|^2 + 4(w - p) \bullet (w - z) = |z - p|^2. \end{aligned}$$

Clearly q is unique. The formulas follow immediately from that for the orthogonal projection given in Lemma 1.3. □

In view of this result we define the *reflexion* of a point z in a line L to be the point $R(z) = 2w - z$ where w is the orthogonal projection of z onto L. Provided z does not lie on L the orthogonal projection w is the mid-point of the line segment joining z, $R(z)$, and L is their orthogonal bisector: when z does lie on L we have $R(z) = z$.

Example 11.1 Let R be reflexion in a line L. It is useful to have the formula of Lemma 11.1 entirely in complex number notation. We keep to the same notation: u is a point on L, T is a unit vector in the direction of L, and N is a unit vector orthogonal to L. We use the fact that for any two vectors a, b (identified with complex numbers) with b unit we have $2(a \bullet b)b = a + \bar{a}b^2$. (Example 1.4.) The reader will now readily verify that the formula for the reflexion can be written as

$$R(z) = u + (\bar{z} - \bar{u})T^2 = u - (\bar{z} - \bar{u})N^2. \qquad (11.2)$$

An immediate consequence of these explicit formulas is that *reflexions are indirect isometries*: indeed, each has the form $R(z) = U\bar{z} + B$ where $U = T^2 = -N^2$, $B = u - \bar{u}T^2 = u + \bar{u}N^2$, and where U is necessarily a unit complex number.

Example 11.2 A special case arises when L is a line through the origin, so we can choose $u = 0$. Then (11.2) reads $R(z) = T^2\bar{z} = -N^2\bar{z}$. We recover the familiar formulas of school geometry for reflexion in a line through the origin as follows. When L makes an angle $\theta/2$ with the x-axis a unit direction T is given by $T = e^{i\theta/2}$, so $T^2 = e^{i\theta}$ and $R(z) = e^{i\theta}\bar{z}$. Writing $Z = R(z)$, $z = x + iy$, $Z = X + iY$ that yields the required formulas

$$X = x\cos\theta + y\sin\theta, \quad Y = x\sin\theta - y\cos\theta.$$

For the purposes of this text it is neither necessary nor desirable to analyze the structure of general indirect isometries. We content ourselves by pointing out the following neat result.

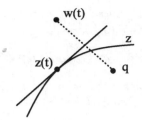

Figure 11.1. Construction of the orthotomic

Lemma 11.2 *Any isometry I whose set of fixed points is a line L is necessarily reflexion in L, and hence indirect.*

Proof Let r be a point with $I(r) \neq r$. As in Lemma 6.4 the fixed point set is the orthogonal bisector of r, $I(r)$. Let R denote reflexion in L, and let p, q be two distinct points on L. Then $J = RI$ is an isometry, and $J(p) = R(p) = p$, $J(q) = R(q) = q$, $J(r) = r$ so p, q, r are non-collinear fixed points of J. It follows immediately from Lemma 6.4 that J is the identity, hence that $I = R$. $\square$

Exercises

11.1.1 Show that any reflexion R has order 2 in the group of isometries, so is self inverse.

11.1.2 Let z be a curve, let L be a line, and let R be reflexion in L. By the *reflexion* of z in L we mean the curve z_L defined by $z_L(t) = R(z(t))$. Let s, s_L be the speeds of z, z_L, and κ, κ_L the curvatures. Show that $s(t) = s_L(t)$ for all t, and that $\kappa(t) = -\kappa_L(t)$ when t is regular.

11.2 Orthotomics

Let z be a regular curve, and let q be a point in the plane. For any parameter t write $w(t)$ for the reflexion of q in the tangent line to z at t. (Figure 11.1.) The resulting curve w is called the *orthotomic* of z with respect to the *source* q. There is of course one orthotomic for each position of the source q, so we have a two parameter family of orthotomics associated to z, one for each choice of q. The formula of the following result follows immediately from the formulas exhibited as (11.1).

Lemma 11.3 *For a regular curve z the orthotomic w with respect to the source q is given by the following formula:*

$$w(t) = (2z(t) - q) + 2\{(q - z(t)) \bullet T(t)\}T(t)$$

where $T(t)$ is the unit tangent vector to z at t. Alternatively, writing $N(t)$ for the unit normal vector, we have

$$w(t) = q - 2\{(q - z(t)) \bullet N(t)\}N(t).$$

It can be useful to have this formula in complex number notation. To that end, the formulas of Lemma 11.1 are replaced by (11.2) to give

$$w(t) = z(t) + \{\overline{q} - \overline{z}(t)\}T(t)^2 = z(t) - \{\overline{q} - \overline{z}(t)\}N(t)^2. \tag{11.3}$$

The family of orthotomic curves associated to even the simplest regular curves provides surprising geometric variety. One general insight is that the source q itself lies on the orthotomic if and only if there is a tangent line passing through q. The remainder of this section is devoted to the surprisingly interesting examples provided by conics, where this point can be vividly illustrated. When the source is 'outside' the conic there are two tangents through it, so we expect the orthotomic to exhibit a self crossing at q: when the source is on the curve itself we expect this self crossing to degenerate to a cusp; and when the source is 'inside' the conic we do not expect the orthotomic to exhibit any self crossings at all.

Example 11.3 The orthotomic of the standard circle $z(t) = e^{it}$ with respect to the source $q = (h, 0)$ where $h > 0$ is easily checked to be the limacon $w(t) = 2e^{it} - he^{2it}$. (Exercise 11.2.1.) In Section 13.3, using a rather different set of ideas, we will discover a very simple reason why the orthotomic of a circle should turn out to be a trochoid. Recall that the form of the limacon depends on the value of h. (Table 7.1.) In particular, for $h > 1$ the limacon is nodal with self crossing at the source; for $h = 1$ the limacon is a cardioid with cusp at the source; and for $h < 1$ it has no self crossings. That confirms the above expectations for the case of a circle. (Figure 11.2.)

The fact that the orthotomics associated to such a simple curve as a circle exhibit such complex geometry raises the question of what happens with other conics. The parabola provides an interesting study pursued in the next few examples.

11.2 Orthotomics

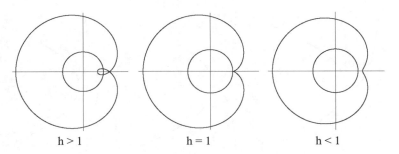

h > 1   h = 1   h < 1

Figure 11.2. Three orthotomics of a circle

Example 11.4 Consider the parabola $x = at^2$, $y = 2at$ with $a > 0$. We will determine the orthotomic with respect to an arbitrary source $q = (\alpha, 0)$ on the x-axis. Applying the formula of Lemma 11.3 we obtain the following parametrization:

$$u(t) = -\alpha + \frac{2(\alpha - a)t^2}{1 + t^2}, \quad v(t) = \frac{2t(\alpha + at^2)}{1 + t^2}.$$

Figure 11.3 illustrates orthotomics for negative, zero and positive values of α. As we expect, the pictures suggest that for $\alpha < 0$ the curve has a loop crossing itself at q: that for $\alpha = 0$ the loop contracts down to a 'cusp' at q: and that for $\alpha > 0$ the loop entirely disappears.

A little analysis throws light on the situation. Let us ask first when the orthotomic of a parabola has irregular parameters.

Example 11.5 The condition for a parameter t to be irregular for the orthotomic is that the equations $u'(t) = 0$, $v'(t) = 0$ have a solution t. A calculation shows that this happens if and only if $\alpha = 0$, in which case $t = 0$ is the *only* irregular parameter, corresponding to a 'cusp' at $q = (0,0)$. (Exercise 11.2.2.) Thus the only case when the orthotomic fails to be regular is when $\alpha = 0$, i.e. the source is the vertex of the parabola, giving rise to the cissoid of Diocles (Example 3.8) parametrized as

$$u(t) = \frac{-2at^2}{1 + t^2}, \quad v(t) = \frac{2at^3}{1 + t^2}.$$

Let us continue our analysis of the orthotomic for a parabola in the case $\alpha \neq 0$ when the orthotomic is regular. To gain further understanding we look for self crossings.

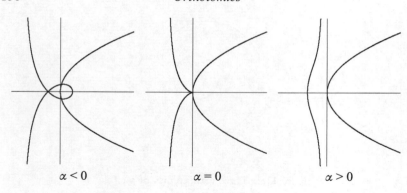

Figure 11.3. Orthotomics of a parabola

Example 11.6 Self crossings appear when there are distinct parameters s, t for which $w(s) = w(t)$, i.e. for which $u(s) = u(t)$ and $v(s) = v(t)$. The condition $u(s) = u(t)$ reduces to $s^2 = t^2$, which since s, t are distinct, is equivalent to $s = -t$. With that substitution, the condition $v(s) = v(t)$ reduces to $t(1 + t^2)(\alpha + at^2) = 0$. The possibility $t = 0$ can be discarded, since then t, $-t$ fail to be distinct. Thus the condition holds if and only if $\alpha + at^2 = 0$, which has a non-zero solution t if and only if $\alpha < 0$. On that assumption the solutions t, $-t$ give rise to a self crossing at the source q, confirming what we expected on the basis of Figure 11.3.

Example 11.7 For positive values of α we obtain a family of orthotomics having no self crossings. An exceptional case arises when $\alpha = a$, i.e. the source is the focus $F = (a, 0)$ of the parabola. In that case the orthotomic is parametrized as $u(t) = -a$, $v(t) = 2at$ with trace the directrix line. Visually, that is entirely consistent with Figure 11.3 since the directrix line clearly represents a transitional case for the orthotomic in the ranges $0 < \alpha < a$ and $a < \alpha$.

Example 11.8 For negative values of α we obtain a family of orthotomics all having a self crossing at the source. An exceptional case arises when the tangents at the self crossing are orthogonal. It is not difficult to verify that this happens if and only $\alpha = -a$, i.e. the source is the point of intersection of the axis and directrix of the parabola. (Exercise 11.2.4.) Translating the source to the origin by writing $U(t) = u(t) + a$, $V(t) = v(t)$ we find that the curve has the following parametrization, which is the

11.2 Orthotomics

right strophoid of Example 2.3.

$$U(t) = 2a\left(\frac{1-t^2}{1+t^2}\right), \quad V(t) = -2at\left(\frac{1-t^2}{1+t^2}\right).$$

The point of the next example is that although general orthotomics of ellipses are very complex curves, they become particularly simple when the source is at a focus.

Example 11.9 Consider the ellipse $x(t) = a\cos t$, $y(t) = b\sin t$ where $0 < b < a$. (Example 3.2.) Recall that the eccentricity is the positive scalar e defined by $a^2 e^2 = a^2 - b^2$, and that the foci are the points $F^- = (-ae, 0)$ and $F^+ = (ae, 0)$. We will determine the orthotomic with respect to the focus F^+. Using the formula of Lemma 11.3 one readily checks that the components u, v of the orthotomic w with respect to F^+ are

$$u(t) = ae - \frac{2ab^2 \cos t \, (e\cos t - 1)}{A}$$

$$v(t) = -\frac{2a^2 b (e\cos t - 1)\sin t}{A}$$

where $A = a^2 \sin 2t + b^2 \cos 2t$. The reader will be forgiven for not immediately recognizing this curve. A rather tedious calculation verifies that u, v satisfy the relation $(u+ae)^2 + v^2 = 4a^2$, so the trace of w is contained in the circle of radius $2a$ centred at the focus F^-. (Figure 11.4.)

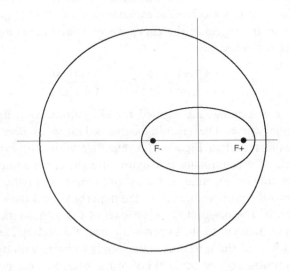

Figure 11.4. Orthotomic of an ellipse with source a focus

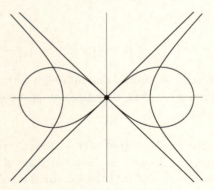

Figure 11.5. Bernoulli's lemniscate

It is natural to conclude this series of examples by asking about the orthotomics of hyperbolas. In general they are very complex curves. A special case, worthy of attention, is provided by the orthotomic of a rectangular hyperbola with respect to its centre.

Example 11.10 The rectangular hyperbola $x^2 - y^2 = a^2$ with $a > 0$ can be 'parametrized' by the formulas $x(t) = a \sec t$, $y(t) = a \tan t$. (Example 3.6.) More precisely, the positive branch is parametrized by restricting to the interval $-\pi/2 < t < \pi/2$, and the negative branch by restricting to $\pi/2 < t < 3\pi/2$. It is a mechanical exercise to verify that the orthotomic with respect to the origin is the curve whose components are given by the following formulas:

$$u(t) = \frac{a \cos t}{1 + \sin^2 t}, \quad v(t) = \frac{-a \sin t \cos t}{1 + \sin^2 t}.$$

Note that these formulas are defined for all parameters t, unlike those for the original curve. The resulting curve is known as *Bernoulli's lemniscate* and illustrated in Figure 11.5, together with the asymptotes of the hyperbola. The lemniscate is a figure-of-eight-curve comprising two (open) loops coming together at the origin. One loop of the lemniscate corresponds to the positive branch of the hyperbola, and the other to the negative branch. The origin itself is *not* part of the orthotomic, reflecting the fact that no tangent to the hyperbola passes through it. However, the limiting positions of the tangents as one tends to infinity along a branch are the asymptotes, so one could think of the source as the point on the orthotomic corresponding to the 'points at infinity' on the hyperbola.

Exercises

11.2.1 Verify that the orthotomic of the circle $z(t) = e^{it}$ with respect to the source $q = (h, 0)$ where $h \geq 0$ is the limacon $w(t) = 2e^{it} - he^{2it}$.

11.2.2 Let w be the orthotomic of the parabola in Example 11.4 with source $(\alpha, 0)$. Verify that when $\alpha = 0$ the only irregular parameter is $t = 0$, and is an ordinary cusp.

11.2.3 Let w be the orthotomic of the parabola in Example 11.4 with source $(\alpha, 0)$. Verify that w passes through the source if and only if $\alpha \leq 0$.

11.2.4 Let w be the orthotomic of the parabola in Example 11.4 with components u, v. Assume that $\alpha < 0$, so the orthotomic has a unique self crossing at the source, and let t be a parameter giving rise to the self crossing. Show that $u'(-t) = -u'(t)$, $v'(-t) = v'(t)$ and deduce that the tangents at the self crossing are orthogonal if and only if $v'(t) = \pm u'(t)$: verify that this condition is equivalent to $\alpha = -a$.

11.2.5 Let w be the orthotomic of the parabola $x(t) = at^2$, $y(t) = 2at$ with respect to a general source $q = (\alpha, \beta)$. Find formulas for the components u, v of w. Verify that in the case when $\beta = 0$ your formulas reduce to those given in Example 11.4.

11.2.6 Show that the orthotomic of the ellipse $x(t) = a\cos t$, $y(t) = b\sin t$ ($0 < b \leq a$) with respect to the origin is the curve w with components

$$u(t) = \frac{2ab^2 \cos t}{a^2 \sin^2 t + b^2 \cos^2 t}, \quad v(t) = \frac{2a^2 b \sin t}{a^2 \sin^2 t + b^2 \cos^2 t}.$$

11.2.7 Show that the orthotomic of the cardioid $z(t) = 2e^{it} - e^{2it}$ with respect to its cusp point is Cayley's sextic. (Example 2.11.)

11.2.8 Let z be a regular curve, and let q be a point. For any parameter t write $w^*(t)$ for the orthogonal projection of q in the tangent line to z at t. The resulting curve w^* is called the *pedal* curve of z with respect to q. Show that the pedal curve is similar to the orthotomic.

11.2.9 Find a formula for the curvature of the orthotomic of a regular curve z with respect to a source q.

11.2.10 Use Lemma 5.1 and the formula for the orthotomic to show that the orthotomic is invariant under parametric equivalence, in the following sense. Let z_1, z_2 be regular curves with domains I_1, I_2, parametrically equivalent via a change of parameter $s : I_2 \to I_1$, let q be a source, and let w_1, w_2 be the orthotomics of z_1, z_2 with respect to q. Show that w_1, w_2 are parametrically equivalent, via the same change of parameter.

11.2.11 Suppose that the curve z_1 and a point q_1 correspond under a congruence C to a curve z_2 and a point q_2. Show that the orthotomic of z_1 with respect to the source q_1 is congruent to the orthotomic of z_2 with respect to the source q_2 under the same congruence C.

11.3 Orthotomics of Non-Regular Curves

We have only defined the orthotomic for *regular* curves. However, it is both natural and fruitful to extend the concept to curves which may exhibit irregular parameters. Let z be a curve, and let q be a point in the plane. Then for any parameter t *at which the limiting tangent line is defined* we can write $w(t)$ for the reflexion of q in the limiting tangent line to z at t, and define the orthotomic to be the resulting 'curve' $w(t)$. At a *regular* parameter t that gives us the same definition of $w(t)$ as before, since by continuity the limiting tangent line coincides with the tangent line. By Lemma 7.10 this definition will also apply at any parameter t for which some derivative of z is non-zero. In a given example we obtain a formula for $w(t)$ from (11.3) in the usual way, and smoothness at irregular parameters for z can be verified by inspection. Bear these points in mind when reading the next example.

Example 11.11 Interesting examples are provided by the orthotomics with respect to the origin of the following epicycloids and hypocycloids. (We exclude the special cases $\lambda = -1, -2$ when the trace collapses to a point, or to an interval.)

$$z(t) = (\lambda + 1)e^{it} - e^{i(\lambda+1)t}.$$

The first thing to note is that z does have irregular parameters. (Example 3.10.) However, they are cusps, so there is a perfectly well defined orthotomic w. (Exercise 7.5.3.) To determine w observe first that

$$z'(t) = 2(\lambda + 1)e^{\left(\frac{\lambda+2}{2}\right)it} \sin \frac{\lambda t}{2}.$$

Thus the unit tangent vector T satisfies $T^2 = e^{(\lambda+2)it}$. In view of the relation (11.3) the orthotomic is given by

$$w(t) = (\lambda + 2)\left\{e^{it} - e^{i(\lambda+1)t}\right\}.$$

11.3 Orthotomics of Non-Regular Curves

Table 11.1. *Orthotomics of epicycloids and hypocycloids*

λ	curve	n	orthotomic
-3	deltoid	3	three leaved clover
-4	astroid	2	four leaved clover
$-5/2$	starfish	5	five leaved clover
1	cardioid	$-1/3$	limacon with $h = 2$

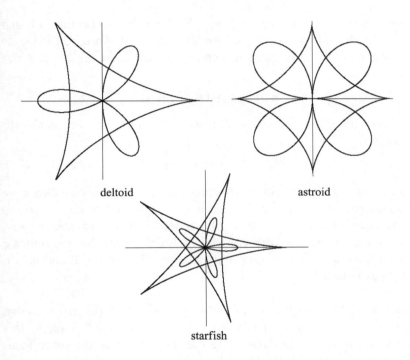

Figure 11.6. Orthotomics of some hypocycloids

In particular w is smooth. Straight calculation now shows that w is parametrically equivalent to a rose curve $2be^{is}\cos ns$, where $b = \lambda+2$ and n is defined by the relation $n = -\lambda/(\lambda+2)$, under the change of parameter

$$t = (n+1)\left(s + \frac{\pi}{n}\right).$$

Some special cases of this example are listed in Table 11.1. In particular, the orthotomics of the deltoid, astroid and starfish with respect to their centres are all clover leaves. (Figure 11.6.)

11.4 Irregular Points on Orthotomics

It is natural to ask which parameters on the curve z give rise to irregular parameters on an orthotomic of z. Here is a satisfying geometric answer.

Lemma 11.4 *Let z be a regular curve, let q be a point not on the trace of z, and let w be the orthotomic of z with respect to q. Then the parameter t is irregular for w if and only if t is inflexional for z.*

Proof Since z is assumed regular we can assume (Lemma 4.1 and Exercise 11.2.10) that it is unit speed. Further, it is no restriction to suppose $q = (0,0)$. According to Lemma 11.3 the orthotomic is given by the formula

$$w(t) = 2(z(t) \bullet N(t))N(t).$$

Differentiating, and using the Frenet-Serret Formulas, we obtain the relation

$$w' = -2\kappa\{(z \bullet N)T + (z \bullet T)N\}$$

where κ denotes the curvature function, and where for convenience we drop the parameter t. Since T, N are linearly independent vectors, $w' = 0$ if and only if $\kappa(z \bullet N) = 0$, $\kappa(z \bullet T) = 0$. However, at least one of $z \bullet N$, $z \bullet T$ must be non-zero: otherwise $z = 0$, which means that the source q lies on the trace of z, contrary to hypothesis. Thus $w' = 0$ if and only if $\kappa = 0$, as required. $\square$

Example 11.12 Recall that the orthotomic of the ellipse $x(t) = a\cos t$, $y(t) = b\sin t$ $(0 < b < a)$ with respect to the focus $F^+ = (ae, 0)$ is a parametrization of the circle of radius $2a$ centred at the other focus $F^- = (-ae, 0)$. (Example 11.9.) Since the ellipse has no inflexions it follows from Lemma 11.4 that this parametrization is regular. We could have verified that fact by direct computation, but it is easier to deduce it from the geometry of the ellipse.

11.5 Antiorthotomics

One application of the envelope construction is to show that the process of taking the orthotomic of a regular curve z can be reversed. Let q be a point, and for each parameter λ on z let z_λ be the orthogonal bisector of the line joining q and $z(\lambda)$: in the case when $q = z(\lambda)$ this line is

11.5 Antiorthotomics

interpreted as the normal line to the curve at λ. Then any envelope of the family of lines z_λ is an *antiorthotomic* for z.

Lemma 11.5 *Let w be a regular curve, and let z be its antiorthotomic with respect to q. Then w is the orthotomic of z with respect to q.*

Proof Write λ for the parameter on w. By definition z is the envelope of the family of lines z_λ, where z_λ is the orthogonal bisector of the line joining q and $w(\lambda)$. Moreover, z_λ is the tangent line to z at the parameter λ. But $w(\lambda)$ is then the reflexion of q in the tangent line to z at λ, so w is the orthotomic of z. □

Example 11.13 Let $F = (a, 0)$ be a fixed point on the x-axis with $a > 0$, and let L be the line with equation $x = -a$. We will show that the antiorthotomic of L, parametrized as $x(t) = -a$, $y(t) = 2\lambda$, is a parabola. First we need a parametrization for the orthogonal bisector z_λ of the line segment joining F to $P = (-a, 2\lambda)$. The mid-point is $Q = (0, \lambda)$. A vector in the direction of FP is $(-2a, 2\lambda)$, and a vector orthogonal to this is $(-2\lambda, -2a)$. Thus the orthogonal bisector z_λ is parametrized as

$$z_\lambda(t) = Z(\lambda, t) = (0, \lambda) + t(-2\lambda, -2a) = (-2t\lambda, \lambda - 2at).$$

For this example $\det J(Z) = \lambda + 2at$, vanishing on the line $\lambda = -2at$ in the (λ, t)-plane, regularly parametrized as $\lambda(u) = -2au$, $t(u) = u$. By Theorem 10.1 that gives rise to the envelope $E(u) = (4au^2, -4au)$ having trace a standard parabola $y^2 = 4ax$ with focus F. Note that L is the directrix of the parabola.

The orthogonal bisectors z_λ in the above example can be traced with ruler and set square: indeed they are the lines through points Q on the y-axis, orthogonal to the lines FQ. Thus we have discovered a practical method for tracing a parabola with given focus F. The reader might like to spend a few minutes tracing a parabola by this method.

Example 11.14 We will determine the antiorthotomic of the standard parabola $x(t) = at^2$, $y(t) = 2at$ where $a > 0$ with respect to its focus $F = (a, 0)$. For convenience, we choose $a = 2$. Consider a general point $P = (2\lambda^2, 4\lambda)$ on the parabola. A calculation similar to that in the previous example shows that the orthogonal bisector z_λ is parametrized as

$$z_\lambda(t) = Z(\lambda, t) = (\lambda^2 + 1, 2\lambda) + t(-4\lambda, 2(\lambda^2 - 1)).$$

We leave the reader to check that $\det J(Z) = 4(1 + \lambda^2)(\lambda + 2t)$, which vanishes if and only if $\lambda = -2t$, a line in the (λ, t)-plane parametrized as $\lambda(u) = u$, $t(u) = -u/2$. Substituting in Z we see that an envelope for Z is given by the following formulas, parametrizing a cubic curve.

$$X(u) = 3u^2 + 1, \quad Y(u) = u(3 - u^2).$$

Exercises

11.5.1 Show that the antiorthotomic of the unit circle $z(\lambda) = e^{i\lambda}$ with respect to $F = (1, 0)$ is the centre $O = (0, 0)$.

12
Caustics by Reflexion

The material of this chapter is drawn from the area of geometric optics, and can be viewed as a class of examples arising naturally in the theory of envelopes. We will be concerned with light caustics in the plane. Here is the idea. One has a mirror (represented by a regular curve z) and a point source of light rays (represented by the lines through a fixed point q): in principle we allow the possibility that q is a 'point at infinity' in order to include the case of a parallel beam of rays. The light rays are reflected off the mirror, and envelop a highly illuminated curve called the 'caustic'. A familiar example is the caustic (Figure 12.1) which appears on the surface of a cup of coffee when sitting in the sunlight. For this example the mirror is represented by a circle (a cross section of the cup) and the source q is the 'point at infinity' on the x-axis.

In Section 12.1 we set up the basic ideas, and in Section 12.2 establish the key result, namely that caustics can be viewed as evolutes of orthotomics. This simple result allows us to discuss the classic example of circle caustics, the consequences of the theory satisfyingly confirmed

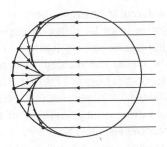

Figure 12.1. Coffee cup caustic

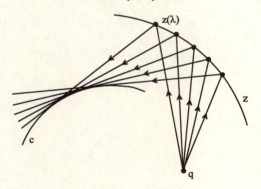

Figure 12.2. Caustic by reflexion

by computer generated pictures of the caustics. Section 12.3 extends the theory to the case when the source is 'at infinity'. Finally, we obtain an alternative view of orthotomics themselves, as envelopes of natural families of circles: that has a useful spin-off, namely a simple method for tracing the orthotomic *by hand*, given a tracing of the curve z.

12.1 Caustics of a Curve

Let z be a regular curve, and let q be a point. Throughout this chapter we refer to z as the *mirror*, to q as the *source*, and write λ for the parameter on z. For each λ there is a unique *incident ray*: when $q \neq z(\lambda)$ it is the line through q, $z(\lambda)$, and when $q = z(\lambda)$ it is the tangent line to z at λ. The *reflected ray* z_λ is the reflexion of the incident ray at λ in the tangent line at λ. (Figure 12.2.) An envelope of the family (z_λ) of reflected rays is called a *caustic* by reflexion of z, with respect to the source q.

Example 12.1 For a circular mirror, there is a ruler and compass construction for tracing caustics by reflexion experimentally. It is based on the observation that for a ray of light, reflected from a circle at X, the points A and A' where the incident and reflected rays meet the circle are equidistant from X. (Figure 12.3.) By drawing a large number of incident rays on a sheet of paper with a ruler, we can then (using only ruler and compass) draw a large number of reflected rays, and the resulting caustic will become visible. The reader is strongly recommended to spend a little while tracing some caustics by this method, to appreciate the simplicity of the construction, and the complexity of the caustic. It

12.1 Caustics of a Curve

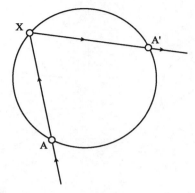

Figure 12.3. Constructing the caustics of a circle

might be interesting to compare the result with the computer generated illustrations in Figure 12.4.

We can pursue the example of a circular mirror in a different way, namely by going back to the basic definitions and computing the envelope explicity. Let us do this for the case when the source is the 'point at infinity' on the x-axis, the 'coffee cup' example.

Example 12.2 Let z be the unit circle $z(\lambda) = e^{i\lambda}$. Consider the pencil of lines parallel to the x-axis. For each parameter λ the unique incident ray through $z(\lambda)$ is reflected in the tangent line at that point. The reflected ray is in the direction $z'(\lambda)^2 = -e^{2i\lambda}$, so can be given parametrically as $z_\lambda(t) = Z(\lambda, t) = e^{i\lambda} - te^{2i\lambda}$. The components of this family of parametrized curves are

$$X(\lambda, t) = \cos\lambda - t\cos 2\lambda, \quad Y(\lambda, t) = \sin\lambda - t\sin 2\lambda$$

and $\det J(Z) = 2t - \cos\lambda$, which vanishes if and only if $2t = \cos\lambda$. By the Envelope Lemma we see that the caustic is the nephroid

$$E(u) = e^{iu} - \frac{1}{2}(\cos u)\, e^{2iu} = \frac{1}{4}\{3e^{iu} - e^{3iu}\}.$$

At first sight there appears to be a discrepancy between the theoretical answer and the illustration of the trace in Figure 12.1. However, a moment's thought should convince the reader that there is no inconsistency. In practice we see only *half* a nephroid, namely the 'real' caustic corresponding to rays reflected from 'inside' the coffee cup: there is also a 'virtual' caustic corresponding to rays reflected from the 'outside'. This

Caustics by Reflexion

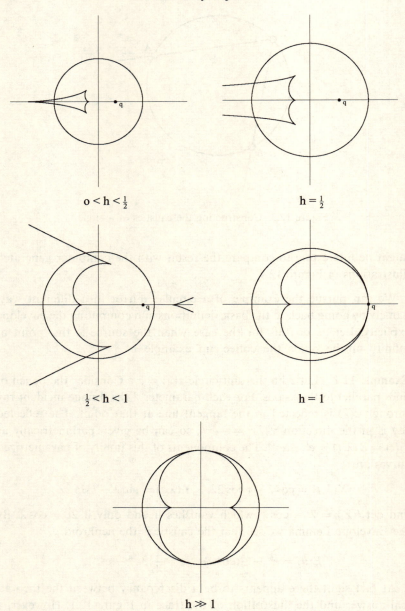

$o < h < \frac{1}{2}$

$h = \frac{1}{2}$

$\frac{1}{2} < h < 1$

$h = 1$

$h \gg 1$

Figure 12.4. Caustics by reflexion for a circle

kind of brutal calculation may well achieve an answer in some examples, but lends little understanding to the overall process. The object of the next section is to establish theoretical results which throw considerable light on the whole business!

Exercises

12.1.1 Consider the caustic of the standard parametrized unit circle $z(t) = e^{it}$ with respect to the source $q = (1,0)$ on the circle. Write down a parametrization of the reflected ray at the point with parameter t, and show (starting from the definitions) that the resulting caustic by reflexion is a cardioid.

12.2 Caustics as Evolutes

The main result of this section is that the caustic of z with respect to a source q is the evolute of the orthotomic of z with respect to q. In this way we can gain valuable geometric information about caustics using our knowledge of evolutes. The next result provides the key technical fact: we deliberately phrase it in greater generality than is immediately necessary, in order to be able to deal with sources 'at infinity' in Section 12.3 of this chapter.

Lemma 12.1 *Let $z, q : I \to \mathbb{R}^2$ be curves with z regular, and let $w : I \to \mathbb{R}^2$ be the curve defined by taking $w(t)$ to be the reflexion of $q(t)$ in the tangent line to z at the parameter t. Then the line joining $z(t)$, $w(t)$ is orthogonal to w at t if and only if the line joining $z(t)$, $q(t)$ is orthogonal to q at the parameter t.*

Proof Note first that the conclusion requires interpretation when $z(t)$, $q(t)$ coincide: in that case $z(t)$, $w(t)$ likewise coincide, and the two lines joining the pairs are each interpreted as the tangent line to z at t. The proof is based on the fact that $z(t)$ is equidistant from $w(t)$, $q(t)$ for all t: symbolically, that is expressed by the identity $|w - z|^2 = |q - z|^2$. Differentiation then yields a further identity

$$(w - z) \bullet (w' - z') = (q - z) \bullet (q' - z')$$

which can be re-written in the form

$$(w - z) \bullet w' = (q - z) \bullet q' + (w - q) \bullet z'.$$

However, $(w - q) \bullet z' = 0$: when $w(t)$, $q(t)$ are distinct that follows from the definition of $w(t)$, and when they coincide it holds trivially. Thus we have the following identity, from which the required result is immediate:

$$(w - z) \bullet w' = (q - z) \bullet q'.$$

$\square$

For the moment we only require the following special case of this result when q is a constant curve, so the line joining $z(t)$, $q(t)$ is automatically orthogonal to q at *any* parameter t. In that case our result reads:

Lemma 12.2 *Let z be a regular curve, and let w be the orthotomic of z with respect to the point q. Then for any parameter t the line joining $z(t)$, $w(t)$ is orthogonal to w at t.*

In Chapter 14 we will see that this result is also a special case of a general one in planar kinematics, representing the key property of 'instantaneous centres' of rotation. (Theorem 14.2.) The principal consequence is the following description of the caustic by reflexion.

Theorem 12.3 *Let z be a regular curve, and let w be the orthotomic of z with respect to a point q. Provided w is regular, and has no inflexions, its evolute is a caustic for z with respect to q.* (The Caustic Theorem.)

Proof Since w is assumed regular, it has a normal line at every parameter t: moreover, by Lemma 12.2 the normal line is the unique line joining the points $z(t)$, $w(t)$. It follows that the normal line to w at t is the reflexion of the line joining $z(t)$, q in the tangent line to z at t, so is the reflected ray to z at t. Any envelope for the normals to w is therefore an envelope for the reflected rays to z, hence (by definition) a caustic by reflexion for z with respect to q. It remains to recall that the envelope of the normals to w is the evolute of w. (Lemma 10.2: we assumed w has no inflexions precisely in order to satisfy the hypotheses of this result.) $\square$

In general we expect only finitely many inflexional parameters on the orthotomic w giving rise to 'points at infinity' on the caustic. However, by deleting these parameter values from the common domain of z, w we can reduce our study to curves for which the orthotomic has no inflexions. Then, in principle the Caustic Theorem enables us to derive the geometry of a caustic from that of the orthotomic. For instance, vertices on the orthotomic will give rise to irregular points on the caustic, with ordinary

12.2 Caustics as Evolutes

vertices corresponding to ordinary cusps. (Lemma 9.2.) These generalities are well illustrated by the caustics of a circle.

Example 12.3 In Example 11.3 we saw that the orthotomics of the standard circle $z(t) = e^{it}$ with respect to the source $q = (h, 0)$, where $h > 0$, are the limacons $w(t) = 2e^{it} - he^{2it}$. The Caustic Theorem tells us that caustics of the circle with respect to q are their evolutes. (Figure 12.4.) The inflexions of limacons were investigated in Example 5.9 and Example 7.7, whilst the vertices were studied in Example 9.6. On this basis we come to the following conclusions.

(i) When $h = 0$ the orthotomic is a circle. The caustic collapses to its centre, namely the source q.
(ii) When $0 < h < 1/2$ the orthotomic is a limacon with four ordinary vertices and no inflexions. The caustic has an arrowhead shape with four ordinary cusps and no points at infinity. As $h \to 0$ the arrowhead contracts down to the origin: and as $h \to 1/2$ so the tip of the arrowhead tends towards the point at infinity on the horizontal axis.
(iii) When $h = 1/2$ the orthotomic is a limacon with three ordinary vertices and one undulation. The caustic has three ordinary cusps and a single point at infinity.
(iv) When $1/2 < h < 1$ the orthotomic is a limacon with four ordinary vertices and two inflexions. The caustic has four ordinary cusps and two points at infinity.
(v) When $h = 1$ the orthotomic is a cardioid with one ordinary vertex, no inflexions, and one ordinary cusp. The caustic has one ordinary cusp, and no points at infinity: indeed the caustic is a cardioid of one third the size. (Table 8.1.)
(vi) When $h > 1$ the orthotomic is a nodal limacon with two ordinary vertices and no inflexions. The caustic has two ordinary cusps, and no points at infinity. Note that the caustic in this case, although rather similar in appearance to a nephroid, does not possess the symmetry of a nephroid, and indeed is not a nephroid. As $h \to \infty$ so the caustic becomes more symmetric, and in the limit does indeed become a nephroid. (Example 12.2.)

The trivial fact, that the caustic of a circle with respect to its centre is simply the centre, appeared as a limiting case in the above example. Here is another example, illustrating the same phenomenon, and extending our knowledge of conics.

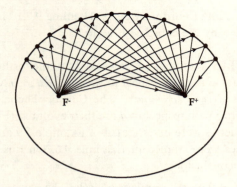

Figure 12.5. The reflective property for an ellipse

Example 12.4 Recall that the orthotomic of the ellipse $x(t) = a\cos t$, $y(t) = b\sin t$ ($0 < b < a$) with respect to the focus $F^+ = (ae, 0)$ is a parametrization of the circle of radius $2a$ centred at the other focus $F^- = (-ae, 0)$. (Example 11.9.) Thus the caustic with respect to F^+ is the evolute of the circle, which is its centre F^-. That establishes a *reflective property* for an ellipse, namely that any ray of light emanating from one focus, and reflected at the point where it meets the ellipse, necessarily passes through the other focus. And that explains why F^- and F^+ are called the 'foci': a beam of light emanating from one, and reflected from the ellipse, 'focuses' at the other. (Figure 12.5.) As $a \to b$ the ellipse tends to a circle, the two foci coalesce to its centre, and we see again the limiting case mentioned above.

Exercises

12.2.1 Determine the caustic of the curve given by $x(t) = 1 - 3t^2$, $y(t) = 3t - t^3$ with the origin as source. Sketch all three curves.

12.2.2 Determine the caustic of the curve given by $z(t) = (\cos t + t\sin t, \sin t - t\cos t)$ with the origin as source.

12.2.3 Consider the equiangular spiral $z(t) = re^{i\gamma t}$ with $r > 0$ and $\gamma = \alpha + i\beta$ where α, β are real and non-zero. (Example 6.6.) Show that the caustic of z with source its pole is another equiangular spiral.

12.2.4 Establish a reflective property for the branches $x(t) = \pm a\cosh t$, $y(t) = b\sinh t$ of a standard hyperbola.

12.3 Sources at Infinity

It is natural to extend the above ideas to the case when the source q is a 'point at infinity', yielding a parallel beam of light. A brutal strategy (in that it produces an answer, but no geometric understanding) is simply to extend the calculation of Example 12.2 to a general curve.

Example 12.5 Consider the caustic of a regular curve z with respect to the pencil of lines parallel to the x-axis. For each parameter λ the unique incident ray through $z(\lambda)$ is reflected in the tangent line at that point. The reflected ray is in the direction $z'(\lambda)^2$. Thus the reflected ray can be parametrized as

$$z_\lambda(t) = Z(\lambda, t) = z(\lambda) + tz'(\lambda)^2.$$

It is now a mechanical task to write down the components of Z and verify that the envelope is the curve with components u, v given by

$$u = x + \frac{y'(x'^2 - y'^2)}{2(x'y'' - x''y')}, \quad v = y + \frac{x'y'^2}{x'y'' - x''y'}.$$

A better strategy is to modify the construction of the orthotomic in such a way that the caustic still turns out to be its evolute. For motivation, return to the case when the mirror is a regular curve z, the source q is a finite point, and the orthotomic of z with respect to q is the curve w constructed in Chapter 11. At this juncture a little thought goes a long way. The key observation is that *the caustic is also the evolute of any parallel w_d of w at distance d*. (Lemma 8.8.) Let us reconsider Figure 11.1 in the light of this comment. At any parameter t the point $w(t)$ is the reflexion of q in the tangent line $L(t)$ at $z(t)$, and the line joining $w(t)$, $z(t)$ is the normal line to w at t. The point $w_d(t)$ lies on this line distant d from $w(t)$: and its reflexion in the tangent line will be a point $q_d(t)$ on the incident ray, distant d from q. Thus the family of parallels w_d corresponds (under reflexion in tangent lines) to the family of parallel circles Q_d of radius d centred at q. The key property of the circles Q_d is that they are orthogonal to the family of incident rays. That provides the geometric clue we are seeking. When q is a 'point at infinity' we should replace the circles Q_d by the family of parallel lines orthogonal to the direction of the incident beam.

So much for motivation: now we proceed formally. Let z be a regular curve, and let Q be any fixed line. We consider the caustic of z with respect to the parallel beam of light orthogonal to Q. For any parameter t write $q(t)$ for the orthogonal projection of $z(t)$ onto Q, and $w(t)$ for the

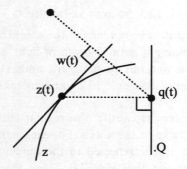

Figure 12.6. Orthotomic with respect to a line

reflexion of $q(t)$ in the tangent line to z at t. The parametrized curve w defined in this way is known as the *orthotomic* of z with respect to the line Q. (Figure 12.6.)

Example 12.6 Let L be the line $y = x$, parametrized as $x(t) = t$, $y(t) = t$. Consider the orthotomic of L with respect to the y-axis Q. Here the tangent line to z at any parameter t is just the line Q itself, the orthogonal projection of $z(t)$ onto Q is $q(t) = (0, t)$, and its reflexion in Q is $w(t) = (t, 0)$. Thus the trace of the orthotomic with respect to the y-axis is the x-axis.

In practice the line Q is determined by one of its points q, and a unit vector v in the direction of the line: then the orthogonal projection $q(t) = q + \{(z(t) - q) \bullet v\}v$, and we obtain a formula for the orthotomic by writing $q(t)$ instead of q in the formula of Lemma 11.3.

Example 12.7 Consider the case when the source is the 'point at infinity' on the x-axis. In that case we can take Q to be the y-axis, $q = (0, 0)$ and $v = (0, 1)$. Write $x(t)$, $y(t)$ for the components of $z(t)$, and $u(t)$, $v(t)$ for the components of $w(t)$. Then $q(t) = (0, y(t))$ and a brief calculation yields the formulas

$$u = \frac{2xy'^2}{x'^2 + y'^2}, \quad v = y - \frac{2xx'y'}{x'^2 + y'^2}. \tag{12.1}$$

In particular, when z is a unit speed curve the formulas simplify to

$$u = 2xy'^2, \quad v = y - 2xx'y'.$$

12.3 Sources at Infinity

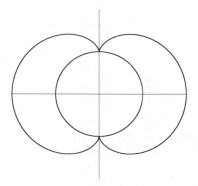

Figure 12.7. Orthotomic of a circle with source at infinity

Example 12.8 Consider the orthotomic of the unit circle $z(t) = e^{it}$ with respect to the y-axis. Using the standard triple angle formulas of trigonometry

$$\cos 3\theta = 4\cos^3 \theta - 3\cos \theta, \quad \sin 3\theta = 3\sin \theta - 4\sin^3 \theta$$

the formula (12.1) produces the curve w defined below. In Example 6.9 we showed that this curve is congruent to a reparametrization of the standard nephroid. (Figure 12.7.)

$$w(t) = \frac{1}{2}\{3e^{it} + e^{3it}\}.$$

Example 12.9 Consider the orthotomic of the parabola $x(t) = at^2$, $y(t) = 2at$ with $a > 0$ with respect to the y-axis. Substituting in the formula for the orthotomic w we find that its components u, v are given by

$$u(t) = \frac{2at^2}{1+t^2}, \quad v(t) = \frac{2at}{1+t^2}.$$

This is a parametrization of the circle of radius a centred at the focus $F = (a, 0)$ with the single point $(2a, 0)$ deleted. (Example 2.6) Observe that the deleted point is the limit of $w(t)$ as $t \to \pm\infty$, so corresponds to the 'point at infinity' on the x-axis.

As in the case of orthotomics with respect to a point, we have the following fundamental relation between the orthotomic with respect to a line, and the caustic with respect to the parallel beam of light orthogonal to that line.

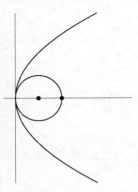

Figure 12.8. Orthotomic of a parabola with source at infinity

Lemma 12.4 *Let z be a regular curve, and let w be the orthotomic of z with respect to a line Q. Provided w is regular, the evolute of w is a caustic for z with respect to the family of lines orthogonal to Q.*

Proof The proof is immediate from Lemma 12.1 since the line joining the orthogonal projection $q(t)$ of $z(t)$ onto Q is automatically orthogonal to Q. □

Example 12.10 By Example 12.8 the caustic of the unit circle $z(t) = e^{it}$ with respect to a beam of light parallel to the x-axis will be the evolute of the nephroid

$$w(t) = \frac{1}{2}\{3e^{it} + e^{3it}\}.$$

However, the evolute of the nephroid is another nephroid of half the size, rotated through a right angle about the origin. (Table 8.1.) Thus the result agrees with that of Example 12.2.

Example 12.11 By Example 12.9 the caustic of the standard parabola $x(t) = at^2$, $y(t) = 2at$ with $a > 0$ with respect to a beam of light parallel to the x-axis will be the evolute of the circle of radius a centred at the focus $F = (a, 0)$, with the single point $(2a, 0)$ deleted. And that is simply the focus F. This establishes a *reflective property* for a parabola, namely that a ray of light parallel to the axis, and reflected at the parabola, necessarily passes through the focus F. Thus a beam of light parallel to the axis, and reflected from the parabola, focuses at F: or put another way, a pencil of rays emanating from F, and reflected from the parabola, produces

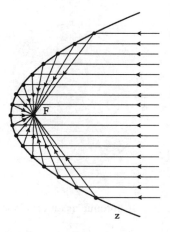

Figure 12.9. The reflective property for a parabola

a beam of light parallel to the axis. (Figure 12.9.) The principle has numerous physical applications, not least the development of parabolic antennas for the propagation of microwave radio signals.

Exercises

12.3.1 Show that the caustic of the parabola $x(t) = 2at$, $y(t) = at^2$ where $a > 0$ with respect to the pencil of lines parallel to the x-axis is the curve with components $u(t) = at(3 - t^2)$, $v(t) = 3at^2$.

12.3.2 Show that the caustic of the natural logarithm curve $x(t) = t$, $y(t) = \log t$ with respect to the pencil of lines parallel to the x-axis is parametrically equivalent to the catenary $u = \cosh(v + 1)$.

12.4 Orthotomics as Envelopes

The practical tracing of orthotomics for a given regular curve z appears to require one to be able to trace a large number of tangents to z. However, there is an alternative description of the orthotomic which obviates this difficulty, and gives rise to a simple practical tracing method. Consider first the case of orthotomics with respect to a finite source q. We claim that *the orthotomic w of z with respect to the source q is an envelope for the family of circles centred at points on z and passing through q*: we have only to observe that by Lemma 12.1 the circle centred at $z(t)$ and passing

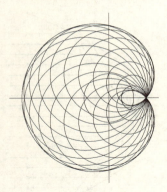

Figure 12.10. Orthotomic as an envelope of circles

through q is tangent to w at t. Thus, given a tracing of z and a source q, one need only trace a large number of circles (using a compass) centred on z and passing through q: the orthotomic will then become visible as the envelope of the circles. We refer to this as the 'circle envelope' tracing method. Figure 12.10 illustrates the method for an orthotomic of a circle. The method extends trivially to the case when the source is a 'point at infinity': thus, by exactly the same reasoning, *the orthotomic w of z with respect to a line Q is an envelope for the family of circles centred at points on z and tangent to Q.*

Of course the above reasoning evades the more subtle question of whether orthotomics are the *only* envelopes produced by the 'circle envelope' method. Indeed that is the case. Here is a formal proof for the case when the source is a point q.

Lemma 12.5 *Let z be a regular parametrized curve. Any envelope of the family of circles through q centred at points on z is either part of the orthotomic with respect to q, or the constant envelope with trace q.*

Proof We use complex number notation. For a fixed value of the parameter λ parametrize the circle centred at $z(\lambda)$ and passing through q in the form

$$Z(\lambda, t) = z(\lambda) + (q - z(\lambda))e^{it}.$$

The partial derivatives with respect to λ and t are then given by the formulas

$$Z_\lambda(\lambda, t) = z'(\lambda)(1 - e^{it}), \quad Z_t(\lambda, t) = (q - z(\lambda))ie^{it}.$$

To determine a pre-envelope for the family we require the points (λ, t) for which the ratio of these complex numbers is a real number α. Set $W = z'(\lambda)/(q - z(\lambda))$: then the required condition is that

$$W\left(\frac{1 - e^{it}}{ie^{it}}\right) = \alpha \quad \text{or equivalently} \quad e^{it} = \frac{W}{W + i\alpha}.$$

Since e^{it} is a *unit* complex number W, $W + i\alpha$ have the same length. Setting $W = U + iV$ with U, V real we see that this happens if and only if $\alpha(\alpha + 2V) = 0$. When $\alpha = 0$ we have $e^{it} = 1$ defining infinitely many 'vertical' lines $t = 2n\pi$ in the (λ, t)-plane, where n is an integer: each such line maps under Z to the point q. When $\alpha = -2V$ we have $W + i\alpha = \overline{W}$ so $e^{it} = W/\overline{W}$: the required envelope is then obtained by substituting for e^{it} in the formula for $Z(\lambda, t)$. A few lines of working produces the envelope in the form

$$Z(\lambda, t) = z(\lambda) + \{\overline{q} - \overline{z}(\lambda)\} T(\lambda)^2$$

where $T(\lambda)$ is the unit tangent vector for z. It only remains to observe that this is the formula for the orthotomic of z with respect to the source q, using complex notation. □

Exercises

12.4.1 Use the 'circle envelope' method to trace the orthotomics of the deltoid and the astroid.

12.4.2 Use the 'circle envelope' method to trace the orthotomics of the unit circle with respect to the source $q = (h, 0)$ for values of h in the following ranges: $h = 0$, $0 < h < 1/2$, $h = 1/2$, $1/2 < h < 1$, $h = 1$, $h > 1$.

12.4.3 Use the 'circle envelope' method to trace the orthotomic of a rectangular hyperbola with respect to its centre. The resulting curve is known as Bernoulli's lemniscate.

13
Planar Kinematics

The main function of this chapter is to provide the reader with an introduction to an important and much neglected area of the physical sciences, namely planar kinematics. It is an area giving rise to substantial illustrations of the ideas developed in this book. Moreover, planar kinematics represents a starting point for spatial kinematics, which will be of considerable future relevance, as robotics assumes a role of ever increasing significance in our daily lives. Some historical background is provided by Section 13.1, centring around a classic example drawn from the engineering literature (the four bar linkage) in which simple mechanical means are used to generate motions of a moving plane. That leads to the abstract concept of a planar motion in Section 13.2, and the associated family of trajectories traced by the points of the moving plane. The concept is illustrated in Section 13.3 by the idea of a general roulette, extending the trochoid construction of Section 3.3 and the involute construction of Section 4.4.

13.1 Historical Genesis

The historical genesis of the subject lies in the Power Revolution, which took place from the thirteenth to the sixteenth centuries. Over that period western man was gradually released from the drudgery of providing a source of power as ways became available of converting water and wind power into mechanical work. Usually, the available power was given by rotary motion (for instance a water wheel or an axle turned by a bullock) and the object was to convert it into linear reciprocating motion (for instance driving a pump to irrigate fields, a bellows for a furnace, or a saw for cutting timber). During the Industrial Revolution mechanisms for converting rotary into linear motion were widely adopted in industrial

13.1 Historical Genesis

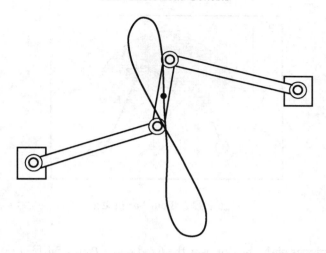

Figure 13.1. The Watt four bar

and mining machinery, locomotives and metering devices. Such devices had to combine engineering simplicity with a high degree of accuracy, and possibly the ability to operate at speed for lengthy periods. For many purposes approximate linear motion is an acceptable substitute for exact linear motion. Perhaps the best known example is the Watt four bar, illustrated in Figure 13.1. The device is made up of three smoothly jointed bars moving with one degree of freedom (dof), the mid-point of the middle (or coupler) bar describing the famous *Watt curve*. The curve has a self crossing with two 'branches' through it, one of which represents an excellent approximation to a straight line. In fact the tangent line has five point contact with the curve at the self crossing, so is a higher undulation.

It was the detailed investigation of the curves traced by planar mechanisms such as the Watt four bar which gave rise to the body of knowledge now known as 'planar kinematics'. The Watt four bar is a special case of the following construction. We have a fixed plane P represented by a sheet of cardboard pinned to a flat surface, and a moving plane Q, represented by a second sheet of cardboard. (The reader is recommended to make a model: the size and shape of the sheets are unimportant.) Two fixed points A and D are marked in P: and likewise two fixed points B and C are marked in Q. Using drawing pins and strips of cardboard (or plastic) we can arrange B to pivot around a circle centred at A, and C to pivot around a circle centred at D. The result is a model in which the

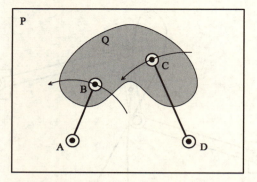

Figure 13.2. Four bar model

plane Q moves with one dof over the fixed plane P in a fairly complicated way: it is called the *four bar linkage*, because there are four *bars* linking the pairs of points AB, BC, CD and DA. (Figure 13.2.) Engineers refer to Q as the *coupler plane* since it couples together the two *cranks* AB, CD. Because of its simplicity and enormous versatility the four bar linkage has a wide range of practical engineering applications.

Suppose now that we mark a fixed *tracing point* W in the moving plane Q. Then W will trace a curve as Q moves over P, called the *trajectory* of W. The trajectory is easily traced on P by the device of making a small hole in Q at W and inserting through it the tip of a coloured pen. A certain amount of experimentation should convince the reader that the trajectories of the planar four bar yield a surprisingly wide range of curves, whose nature and complexity depend on the lengths AB, BC, CD, DA, and on the position of W in the moving plane. An example is illustrated in Figure 13.3.

The complexity of the four bar motion depends on the *bar lengths*, i.e. the distances $a = AB$, $b = BC$, $c = CD$, $d = DA$. We do not require a full analysis of the four bar linkage, but the salient features are worthy of mention. For a given position of B there are two possible positions for C, each the reflexion of the other in the diagonal BD. There may be one motion of the linkage encompassing both positions, or there may be two distinct motions – it all depends on the bar lengths. Writing a, b, c, d in increasing order of magnitude as a', b', c', d' it can be shown that the linkage describes a single motion when $a' + d' > b' + c'$ (the *Grashof type*) and two distinct motions when $a' + d' < b' + c'$. For the Grashof type the trajectory has a single branch, whilst for the non-Grashof type it

13.1 Historical Genesis

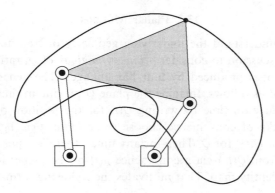

Figure 13.3. A four bar trajectory

has two distinct branches. The boundary between the two types is given by $a' + d' = b' + c'$, i.e. the shortest plus the longest length is the sum of the other two. More generally, the linkage is *collapsible* when the bar lengths satisfy a relation of the form $\pm a \pm b \pm c \pm d = 0$ or, put another way, the points A, B, C, D become collinear at some instant during the motion. A collapsible linkage will normally collapse in just one way. Figure 13.4 illustrates some very special linkages which collapse in more than one way: when $a' = b'$ and $c' = d'$ the linkage collapses in two different ways (the parallelogram, crossed parallelogram and kite) whilst when $a' = b' = c' = d'$ it collapses in three different ways (the rhombus).

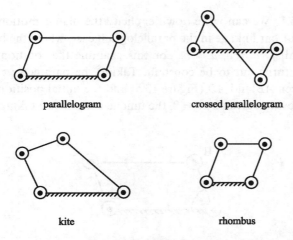

Figure 13.4. Collapsible four bar linkages

13.2 Planar Motions

In order to understand the underlying geometry of trajectories we will widen our discussion to consider *arbitrary motions* of Q, rather than the very special ones produced by four bar linkages. The general idea can be introduced as follows. Imagine the plane Q moving in time t: indicate that dependence on time by writing $Q(t)$ for the position of Q at time t. For the sake of convenience choose some *initial time* t_0 yielding an *initial position* $Q(t_0)$ for Q. Then for any time t there is a position $Q(t)$ of Q obtained from $Q(t_0)$ via a congruence $\mu(t)$ of the plane: in particular $\mu(t_0)$ is the identity map. That motivates the following formalities.

As in Section 6.2 write $SE(2)$ for the group of congruences of the Euclidean plane. By a *planar motion* we mean a 'smooth' mapping $\mu : I \to SE(2)$, where I is an open interval, such that for some parameter t_0 the congruence $\mu(t_0)$ is the identity map. One thinks of μ as a one parameter family $\mu(t)$ of congruences, so really as a 'parametrized curve' *in the group* $SE(2)$ rather than just in the plane. The meaning of the term 'smooth' in this context is as follows. Using complex notation, we can write $\mu(t)(w) = \rho(t)w + \tau(t)$ where $\rho(t)$, $\tau(t)$ are complex numbers and $\rho(t)$ has unit modulus: $\rho(t)$ is the *rotational part* and $\tau(t)$ the *translational part* of the motion. We say that μ is *smooth* when $\rho(t)$ and $\tau(t)$ are smooth functions of t. Given a rigid motion μ and a point w in the plane, the parametrized curve defined by the formula $t \mapsto \mu(t)(w)$ is called the *trajectory* of w under μ.

Example 13.1 We can write down explicity the planar motion μ arising from the four bar linkage in the parallelogram case, where the bar lengths a, b, c, d satisfy $c = a$, $d = b$. For this example the rotational part of the motion turns out to be constant. Taking the parameter t to be the angle between AB and AD (Figure 13.5) and the initial position to be the collapsed position given by $t = 0$, the unique congruence taking $B(0) = a$,

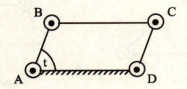

Figure 13.5. The parallelogram four bar

13.2 Planar Motions

$C(0) = a+b$ to $B(t) = ae^{it}$, $C(t) = ae^{it} + b$ is the translation $\mu(t)$ given by

$$\mu(t)(w) = (w-a) + ae^{it}$$

representing a regular parametrization of the circle of radius a centred at the point $(w-a)$. Thus the rotational and translational parts of μ are respectively $\rho(t) = 1$, $\tau(t) = a(e^{it} - 1)$.

For general four bars it is simply not practicable to write down explicit formulas for the motions. Indeed the four bar motion is an extremely subtle example whose mathematics (at the time of writing) is still not fully understood. However, in partial compensation for this disappointment we will consider a limiting case of the four bar motion (also of engineering importance) where the motion can be written down explicitly.

Example 13.2 For the general planar four bar linkage the points B, C in the moving plane Q are restricted to move along *circles* in the fixed plane P. If we think of a line as a limiting case of a circle (tangent to that line at a fixed point with radius tending to infinity) then a limiting case of the planar four bar is obtained when B, C are restricted to move along *lines* in the fixed plane P, as illustrated in Figure 13.6. Mechanical engineers refer to this motion as the *double slider*. For the double slider it is rather easy to write down a formula for μ. For simplicity take the lines to be the x-axis and the y-axis, the distance between B, C to be 1, and the parameter t to be the angle between the x-axis and the line BC. For general t we have $B(t) = i \sin t$ and $C(t) = \cos t$. Suppose we take $t = 0$ to be the initial parameter, for which $B(0) = 0$, $C(0) = 1$. Then we seek a congruence $\mu(t)(w) = \rho(t)w + \tau(t)$ which takes $B(0)$ to $B(t)$ and $C(0)$ to $C(t)$: the reader is encouraged to check that that gives

$$\rho(t) = \cos t - i \sin t, \quad \tau(t) = i \sin t.$$

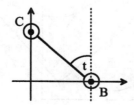

Figure 13.6. The double slider

Exercises

13.2.1 In Example 13.2 an explicit formula is derived for the double slider motion. Writing the tracing point as $w = (u, v)$ show that the trajectory generated by w is given parametrically as

$$x(t) = -v \sin t + u \cos t, \qquad y(t) = (u+1) \sin t + v \cos t.$$

Deduce that the trace of the trajectory is an ellipse, possibly degenerating to a line segment.

13.3 General Roulettes

The next step in our discussion is to pursue the idea of a roulette rather further. There is considerable gain in replacing the circles of Section 3.3 by general regular curves: by so doing we will see that familiar classes of curves turn out to be roulettes, lending cohesion to the subject.

Consider two parametrized curves p, q having the same domain I. Think of p, q lying in superimposed planes P, Q at time t_0: P is thought of as the *fixed plane*, and Q as the *moving plane*. Correspondingly we think of p as the *fixed curve* and q as the *moving curve*. A simple physical model can be made by taking P, Q to be plastic transparencies, and tracing p on P and q on Q with differently coloured pens. The curve q is then rolled along the curve p, carrying with it the plane Q until at time t we have the situation illustrated in Figure 13.7. In the moving plane we

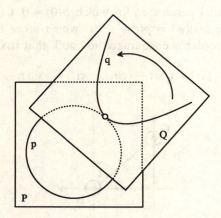

Figure 13.7. The idea of a roulette

13.3 General Roulettes

choose any *tracing point* w, fixed relative to the moving curve q. Then, as the moving plane Q moves, so w traces out the required roulette.

We need to interpret the physical idea of 'q rolling on p without slipping'. To this end we assume henceforth that p, q *have the same speeds for all parameters* t, i.e. that $|p'(t)| = |q'(t)|$ for all t. It is an immediate consequence of this assumption that for any fixed parameter t_0 the arc lengths of p, q from t_0 to t are equal for all t: and that is the required interpretation. It will be convenient to say that p, q are *equitangent* at the parameter t when $p(t) = q(t)$ and $p'(t) = q'(t)$. The basic mental picture of the preceding paragraph is that at any instant t, the curve q in the moving plane gives rise to a curve q_t in the fixed plane, equitangent to p at t. Here is the formal expression of this intuition.

Lemma 13.1 *Let p, q be regular curves with the same domain having equal speeds. Then for any choice of parameter t there exists a (unique) congruence μ for which the curves p, $q_t = \mu(t) \circ q$ are equitangent at t. The rotational and translational parts ρ, τ of μ are given by*

$$\rho = \frac{p'}{q'}, \quad \tau = \frac{pq' - p'q}{q'}. \tag{13.1}$$

Proof The conditions for p, q_t to be equitangent at t are that $p(t) = \mu(t)(q(t))$ and $p'(t) = \rho(t)q'(t)$. The latter formula defines $\rho(t)$ uniquely by $\rho(t) = p'(t)/q'(t)$. Since p, q have the same speeds $\rho(t)$ is a unit complex number, so represents a rotation. The former condition then defines $\tau(t)$ uniquely as $\tau(t) = p(t) - \rho(t)q(t)$. $\square$

Note that this proof does not assume that there exists an 'initial' parameter t_0 for which p, q are equitangent, though in all our examples that will be the case. We define the *roulette* traced by the point w to be the planar motion $t \mapsto \mu(t)$. The practical value of Lemma 13.1 is that given a fixed tracing point w, we then have a useful parametrization for the associated trajectory $t \mapsto \mu(t)(w)$. Indeed by that result we have

$$\mu(t)(w) = p(t) + \frac{p'(t)}{q'(t)}\{w - q(t)\}. \tag{13.2}$$

In the next two examples note that the parametrizations p, q have been set up to ensure that they have the same speed, and that there is an initial parameter ($t = 0$ in both examples) at which p, q are equitangent.

Example 13.3 Let us derive again the parametrization of a trochoid in Section 3.3. Recall that C is the pitch circle, with centre O at the origin, and radius $R > 0$: and C' is the rolling circle, with centre O', and radius R'. (Figure 3.7.) Parametrize the pitch and rolling circles as

$$p(t) = Re^{it}, \quad q(t) = (R + R') - R'e^{-Rit/R'}.$$

The reader will readily check that p, q are equitangent at $t = 0$ and have the same constant speed R. We take the tracing point to be the point $w = (R+R') - hR'$, so in the initial configuration w is the point on the line joining the centres of C, C' distance hR' from the centre of C'. A line or two of calculation using (13.2) now yields the required parametrization, agreeing with (3.1),

$$z(t) = (R + R')e^{it} - hR'e^{i(\frac{R+R'}{R'})t}.$$

Example 13.4 Consider again the roulette of a tracing point w carried by a circle C of radius $R > 0$ rolling along a straight line L. (Example 3.15.) We take L to be the x-axis. It is assumed that in the initial configuration C is the circle of radius R centred at the point iR, and that $w = iR(1 - h)$ is the point on the y-axis distance hR from the centre of C, where $h > 0$. (Figure 3.10.) Then C is naturally parametrized as $q(t) = iR - iRe^{it}$, and L as $p(t) = Rt$. Clearly p, q are equitangent at $t = 0$, and have the same constant speed R. The formula of Lemma 13.1 yields the same parametrization as Example 3.15, namely

$$x(t) = R(t - h\sin t), \quad y(t) = R(1 - h\cos t).$$

Example 13.5 *In which we show that involutes can be viewed as roulettes.* Let p be a unit speed curve, and let t_0 be a fixed parameter. By a reparametrization we can suppose that $t_0 = 0$, and by applying a congruence we can suppose that $p(0) = 0$, $p'(0) = 1$. Parametrize the tangent line at t_0 as $q(t) = t$. Then p, q have the same speeds, and are equitangent at $t = 0$. The formula for the trajectory of w then yields

$$\mu(w)(t) = p(t) + \frac{p'(t)}{q'(t)}\{w - q(t)\} = p(t) - (t - w)p'(t).$$

Looking at the formula for the involute of a regular curve p in Section 4.4 we see that *involutes of regular curves p are trajectories of the roulettes obtained by rolling a tangent line q on p*.

Example 13.6 *In which we show that orthotomics can be viewed as roulettes.* Let p be a regular curve, and let t_0 be a fixed parameter. By reparametrization we can suppose $t_0 = 0$, and by applying a congruence that $p(0) = 0$,

$p'(0) = 1$. Define a curve q by taking $q(t) = \overline{p(t)}$, the reflexion of $p(t)$ in the tangent line to p at $t = 0$. (Figure 12.7.) Clearly p, q have the same speeds, and are equitangent at $t = 0$. According to (13.2) the resulting roulette is given by the formula

$$\mu(w)(t) = p(t) + \frac{p'(t)}{q'_t(t)}\{w - q(t)\} = p(t) + \{w - \overline{p}(t)\}T(t)^2$$

where $T(t)$ is the unit tangent vector to p. Looking at Lemma 11.3 we see that this is the orthotomic of z with respect to the source $\overline{w}$. Thus *orthotomics of regular curves p can be characterized as trajectories of the roulettes obtained by rolling a reflexion of p in some tangent line on p.* In particular, this helps us to understand why the orthotomics of a circle are limacons. (Example 11.3.)

Exercises

13.3.1 Let $a > 0$. Consider the roulette generated by the fixed parabola $x(t) = at^2$, $y(t) = 2at$ and a moving parabola $x(t) = -at^2$, $y(t) = 2at$. Show that the roulette generated by the vertex of the moving parabola is similar to the cissoid of Diocles. (Example 11.4.) That should come as no surprise. According to Example 13.6 the roulette is the orthotomic of the fixed parabola with respect to its vertex, which was shown in Example 11.4 to be the cissoid of Diocles.

13.3.2 Let p_1, q_1 be regular curves with domain I and equal speeds; and let p_2, q_2 be the curves with domain J obtained via a change of parameter $s : J \to I$. (Thus p_2, q_2 are likewise regular, and have equal speeds.) Let μ_1, μ_2 be the roulettes obtained by rolling q_1, q_2 on p_1, p_2. Show the trajectories of any tracing point w under μ_1, μ_2 are parametrically equivalent under the *same* change of parameter s. (In view of Example 13.6 this generalizes the result of Exercise 11.2.10.)

14
Centrodes

In this chapter we will gain some understanding of the nature of a general planar motion μ via two very important curves associated to μ. The first real illumination arises by asking, at any given instant t, for those tracing points w with the property that t is irregular for the trajectory under w. In general, the answer is that there is a *unique* tracing point w with this property, the 'instantaneous centre' of rotation at that instant. That leads naturally to the (fixed and moving) centrodes associated to general motions, and to the classical result of Chasles, that such motions arise as the roulettes associated to these two curves. That provides the content of Section 14.3. In this way the concept of a roulette finally sheds its mantle as an amusing construct for special curves, and assumes its central role as a significant geometric idea in planar kinematics. This basic result provides more than just an insight into the nature of planar motions: it allows one to deduce useful properties of the motion from the geometry of the centrodes.

14.1 Generic Parameters

Recall that for a planar motion μ the trajectory generated by the tracing point w can be written in the complex form $\mu(t)(w) = \rho(t)w + \tau(t)$ where $\rho(t)$, $\tau(t)$ are complex numbers and $\rho(t)$ has unit modulus. Given a parameter t it is natural to ask for which tracing points w the parameter t is irregular for the trajectory. Although it has little technical content, the next result underpins the whole development of planar kinematics.

Lemma 14.1 *Let μ be a planar motion. Then, for each parameter t with $\rho'(t) \neq 0$ there is a unique tracing point $w = q(t)$ for which t is irregular*

14.1 Generic Parameters

for the trajectory of w, given by the formula $q(t) = -\tau'(t)/\rho'(t)$ *where* ρ, τ *are the rotational and translational parts of* μ.

Proof The condition on w is that the derivative of the trajectory should vanish, i.e. that $0 = \rho'(t)w + \tau'(t)$, which has a unique solution for w provided $\rho'(t) \neq 0$. $\square$

This result motivates the following definition. The parameter t is *generic* for the motion μ when $\rho'(t) \neq 0$. Put another way, a parameter t fails to be generic when the rotational part has a stationary value. One expects only a discrete set of parameters t to fail to be generic; however, that is not necessarily the case, as the next example shows.

Example 14.1 Consider the planar motion μ of Example 13.1 arising from the four bar linkage in the parallelogram case, where the bar lengths a, b, c, d satisfy $c = a$, $d = b$. Taking the parameter t to be the angle between AB and AD, and the initial position to be the collapsed position given by $t = 0$, we saw that the rotational and translational parts of μ are given by $\rho(t) = 1$, $\tau(t) = a(e^{it} - 1)$: thus the rotational part is constant, and *every* parameter t fails to be generic.

The next result is central to planar kinematics: that at any generic instant t all the normals to the trajectories pass through the single point $p(t) = \mu(t)(q(t))$ in the fixed plane.

Theorem 14.2 *Let t be a generic parameter for a motion μ: then for any tracing point $w \neq q(t)$ the normal line at t to the trajectory $\mu(t)(w)$ passes through $p(t)$.* (Theorem of the Instantaneous Centre.)

Proof Under the hypothesis of the theorem the trajectory associated to w is regular at t, so the tangent and normal lines at t are defined. Write $z(t) = \mu(t)(w) = \rho(t)w + \tau(t)$ with $\rho(t)$ a rotation and $\tau(t)$ a translation. In the following we drop the parameter t, for notational convenience. Recall first that the normal line at t to the trajectory is the set of points r for which $(r - z) \bullet z' = 0$. We have to show that this relation holds for $r = p$, i.e. that $(p - z) \bullet z' = 0$. Note first that

$$(z - p) \bullet (z - p) = \rho(w - q) \bullet \rho(w - q) = (w - q) \bullet (w - q).$$

Differentiating both sides with respect to the parameter t we obtain

$$(z - p) \bullet (z' - p') = (w - q) \bullet -q'.$$

Note that $p' = \rho(q')$: indeed, differentiating the identity $p = \rho q + \tau$ we obtain $p' = \rho q' + \rho' q + \tau' = \rho q'$, since $\rho' q + \tau' = 0$. That gives

$$\begin{aligned}(z-p) \bullet z' &= (z-p) \bullet p' - (w-q) \bullet q' \\ &= \rho(w-q) \bullet \rho(q') - (w-q) \bullet q' \\ &= (w-q) \bullet q' - (w-q) \bullet q' \\ &= 0.\end{aligned}$$

□

The crucial intuition behind this result is that the moving plane is rotating 'instantaneously' about the point $p(t)$. For this reason, the unique tracing point $q(t)$ of Lemma 14.1 is called the *moving instantaneous centre of rotation* at time t, and the corresponding point $p(t) = \mu(t)(q(t))$ is the *fixed instantaneous centre*. Thus $p(t)$, $q(t)$ are defined by the formulas

$$q(t) = -\frac{\tau'(t)}{\rho'(t)}, \quad p(t) = \frac{\rho'(t)\tau(t) - \rho(t)\tau'(t)}{\rho'(t)}. \tag{14.1}$$

14.2 Generic Parameters for Roulettes

We gain geometric insight into the nature of generic parameters, when we consider the class of roulettes. (As we will see, that is not as restrictive an assumption as one might imagine it to be.) There is an underlying intuition here, worthy of detailed explanation. Consider the roulette μ obtained by rolling one curve q on another curve p. For a given parameter t we can think of p, q being approximated by their circles of curvature C_p, C_q of radii R_p, R_q. The intuition is that in some sense μ will be approximated at t by the roulette arising by rolling the circle C_q on the circle C_p. (That is another reason why we isolated the study of trochoids in Chapter 3.) Further, one expects the irregular points for the trajectories to be related to those for the associated trochoids. Let us recall the situation for trochoids.

Example 14.2 Write λ for the ratio R_p/R_q of the radii. There are two special cases. The first is when $\lambda = -1$, corresponding to the relation $\kappa_p = \kappa_q$. In that case the circle C_p has the same radius as C_q and is *inside* it: thus no rolling takes place, so the trace of the trochoid is a point, and *every* parameter t is irregular. However, when $\lambda \neq -1$, irregular parameters arise if and only if the tracing point is on the circumference of C_p. (Example 3.10.) Such parameters are necessarily cusps, failing to be ordinary if and only if $\lambda = -2$, corresponding to the relation $2\kappa_p = \kappa_q$.

14.2 Generic Parameters for Roulettes

(Exercise 7.5.3.) Geometrically, the latter case occurs when C_q is a circle of half the radius of C_p and rolling *inside* it: thus we have Cardan circles where the general trajectory is an ellipse, collapsing to a diameter of C_q for points on the circumference of C_p.

Bearing this guiding intuition in mind, the next result makes considerable sense.

Lemma 14.3 *Let p, q be regular curves with the same domain having equal speeds, let μ be the roulette obtained by rolling q on p, and let t_0 be a parameter for which p, q are equitangent. Then t fails to be generic if and only if the curvatures κ_p, κ_q of p, q at t_0 are equal.*

Proof It follows from the proof of Lemma 4.1 that we can assume p, q are unit speed curves under the *same* change of parameter, and that $t_0 = 0$. Further, by applying a suitable congruence we can assume that $p(0) = q(0) = 0$, and that $p'(0) = q'(0) = 1$. By Lemma 13.1 the rotational part of μ is given by $\rho = p'/q'$, and 0 fails to be generic if and only if $0 = \rho'(0)$, i.e. if and only if

$$0 = p''(0)q'(0) - p'(0)q''(0) = p''(0) - q''(0).$$

Since p has unit speed $p' \bullet p' = 1$ identically. Differentiating, we see that $p' \bullet p'' = 0$ identically, hence that $p'(0) \bullet p''(0) = 0$, equivalent to $\Re p''(0) = 0$. Likewise $\Re q''(0) = 0$. Thus $p''(0) = q''(0)$ if and only if $\Im p''(0) = \Im q''(0)$. By Exercise 5.3.8 that is equivalent to $\kappa_p = \kappa_q$. □

Example 14.3 In Example 13.5 we saw that involutes of a regular curve p could be considered as the trajectories of the roulette obtained by rolling one of the tangent lines q along the curve. It is illuminating to ask for the condition on a parameter t to be non-generic. By Lemma 14.3 the condition is that $\kappa_p(t) = \kappa_q(t)$: now $\kappa_q(t) = 0$, since q is a line, so the condition is that $\kappa_p(t) = 0$, i.e. that t *is inflexional* for p. Intuitively, that is not too surprising: as one moves through an inflexion one expects the moving plane containing the tangent line to change its sense of rotation, so that no matter which tracing point we choose there should be a visible irregular point on the trajectory.

Example 14.4 In Example 13.6 we considered the roulette generated by regular curves p, q where q is the reflexion of p in one of its tangent lines. The resulting trajectories are then the orthotomics of p. By Lemma 14.3

the condition for a parameter t to be non-generic is that $\kappa_p(t) = \kappa_q(t)$. However, the curvatures are related by $\kappa_p(t) = -\kappa_q(t)$, so the required condition is that $\kappa_p(t) = \kappa_q(t) = 0$, i.e. that t *should be inflexional for* p. This example shows that if we start with a mirror p having no inflexions, and take a light source not on the mirror, then all the resulting orthotomics are regular curves.

14.3 Fixed and Moving Centrodes

Let μ be a planar motion with rotational and translational parts ρ, τ. We say that μ is *generic* when every parameter t is generic. We keep to the notation of the previous chapter, writing $p(t)$, $q(t)$ for the fixed and moving instantaneous centres at the instant t. The curve $q(t)$ is called the *moving centrode* of μ; it is thus the locus of all moving instantaneous centres of rotation. Likewise, the curve $p(t)$ is the *fixed centrode* of μ, and is the locus of all fixed instantaneous centres of rotation. The following examples illustrate the mechanics of determining centrodes for some of the motions introduced in the previous chapter.

Example 14.5 For the motion μ arising from the double slider of Example 13.2 the rotational part is $\rho(t) = \cos t - i \sin t$, and the translational part is $\tau(t) = i \sin t$. Clearly, μ is generic. The moving centrode is

$$q(t) = -\frac{\tau'(t)}{\rho'(t)} = \cos t(\cos t + i \sin t).$$

The trace of $q(t)$ is a circle: the components $x(t) = \cos^2 t$, $y(t) = \sin t \cos t$ satisfy the equation $x^2 + y^2 = x$ of the circle radius $1/2$ and centre $(1/2, 0)$. The fixed centrode is

$$p(t) = \rho(t)q(t) + \tau(t) = \cos t + i \sin t$$

parametrizing the unit circle $x^2 + y^2 = 1$. Thus the traces of the centrodes are circles, the fixed circle of twice the radius of the moving circle, and tangent to it at the point $(1, 0)$. (Figure 14.1.)

A consequence of Theorem 14.2 is that in principle the fixed instantaneous centre is determined once we know the normals of any *two* trajectories at time t: indeed it must be the intersection of the normals. This remark can sometimes be used to locate the instantaneous centre geometrically, even when a formula for the motion μ is not known.

14.3 Fixed and Moving Centrodes

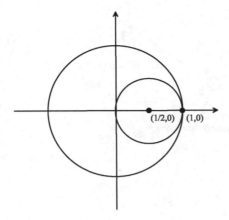

Figure 14.1. Centrodes of the double slider

Example 14.6 Consider again the double slider of Example 13.2. For this example it is clear that the trajectory of B is the 'vertical' diameter of the unit circle, and that the trajectory of C is its 'horizontal' diameter. (Note that the irregular points of these trajectories are the extremities of the diameters.) Thus the tangent line to the trajectory of B is the y-axis, and the normal line is the 'horizontal' line through B: likewise, the tangent line to the trajectory of C is the x-axis, and the normal line is the 'vertical' line through C. It follows that the instantaneous centre is the point of intersection of the 'horizontal' line through B with the 'vertical' line through C: elementary geometry should now convince the reader that the trace of the fixed centrode is therefore the unit circle, agreeing with the result of the computation in Example 14.5.

In principle we can identify the fixed centrode q for a planar four bar linkage. (We neglect here the question of whether parameters are generic.) Indeed, the trajectories of B, C are circles centred at A, D, the normal lines to the trajectories are the lines AB, CD, and generally these will intersect at the fixed instantaneous centre I. In this way we obtain a range of fascinating centrode curves, whose form depends crucially on the bar lengths a, b, c, d. An analysis of the general centrode lies well beyond an undergraduate text: however, there are special cases where elementary geometry suffices.

Example 14.7 Explicit illustrations are provided by crossed parallelogram four bars. According to the preceding example the trace of the fixed

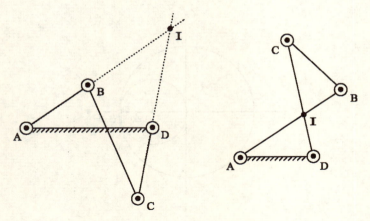

Figure 14.2. Crossed parallelogram four bars

centrode is the locus of the intersection I of the diagonals AB and CD. For $b < a$ the point I is *within* those line segments: but for $b > a$ it lies *outside* them. (Figure 14.2.) In either case the symmetry of the diagram tells us that $DI = IB$. Thus when $b < a$ we have $AI + ID = AI + IB = AB = a$, and by Exercise 3.1.7 the trace is contained in an ellipse with foci A, D: however when $b > a$ we have $AI - ID = (AB + BI) - ID = AB = a$, and by Exercise 3.1.8 the trace is contained in a hyperbola with foci A, D. Figure 14.3 illustrates the ellipse case.

We can widen our range of examples considerably by identifying the centrodes for the class of motions provided by roulettes.

Lemma 14.4 *Let p, q be regular curves having equal speeds, and let μ be the roulette obtained by rolling q on p. Assume that μ is generic. Then the fixed and moving centrodes P, Q associated to μ coincide with p, q.*

Proof By Lemma 13.1 we have $\mu(w) = \rho w + \tau$ (where we drop the parameter t) with $\rho = p'/q'$, $\tau = (pq' - p'q)/q'$. Differentiation yields $\tau' = -q\rho'$, so the moving centrode is $Q = -\tau'/\rho' = q$, and the fixed centrode is $P = \rho Q + \tau = \rho q + \tau = p$. □

In particular, when every parameter is generic, the only tracing points giving rise to trajectories having irregular parameters are those on the moving curve q. The reader may like to correlate that statement with the examples of trochoids discussed in Chapter 3 where cusps on the trajec-

14.3 Fixed and Moving Centrodes

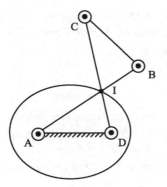

Figure 14.3. Centrode for a crossed parallelogram

tories corresponded to tracing points on the rolling circle, the cardioid providing a good illustration.

The final result of this chapter encapsulates the reason for introducing centrodes in the first place. It tells us that *under certain generality conditions* any planar motion is a roulette. The practical import of this result is that properties of general motions μ can be deduced from the geometry of their centrode curves p, q.

Theorem 14.5 *Let μ be a generic motion whose fixed and moving centrodes p, q are regular. Then p, q have the same speed. Also, μ coincides with the roulette μ^* obtained by rolling q on p.* (Chasles' Theorem.)

Proof Write ρ, τ for the rotational and translational parts of μ. The centrodes p, q are then expressed in terms of ρ, τ by (14.1). Differentiating these expressions we obtain $p' = \rho q'$: then, taking moduli of both sides we find that p, q have equal speeds. Thus the roulette μ^* obtained by rolling q on p is well defined, and its rotational and translational parts ρ^*, τ^* are determined by (13.1). A few lines of working verify that $\rho^* = \rho$, $\tau^* = \tau$. The result follows. □

Example 14.8 By Example 14.5 the traces of the centrodes for the double slider are circles, one of twice the radius of the other, and tangent at a point. These are the Cardan circles of Example 3.12 where the trajectories of the resulting roulette are ellipses, degenerating to a diameter of the fixed circle when the tracing point is on the moving circle. By the above result, the trajectories of the double slider are likewise ellipses, possibly degenerating to a diameter of the fixed circle. (Exercise 13.2.1 established this fact by direct computation.)

Exercises

14.3.1 Let μ be a planar motion. The *inverse motion* μ^{-1} is defined by $\mu^{-1}(t) = \mu(t)^{-1}$. Suppose μ is generic. Show that the moving centrode for μ^{-1} is the fixed centrode for μ, and that dually the fixed centrode for μ^{-1} is the moving centrode for μ.

15
Geometry of Trajectories

In this chapter we will study how exceptional geometric features (irregular points, inflexions and vertices) appear on the trajectories of a general planar motion. To put matters into perspective it helps to make some preliminary comments on the exceptional features which can appear on a general curve z. The first thing to notice is that irregular points are basically different from inflexions and vertices in that they represent the common zeros of *two* smooth functions in one variable (the derivatives of the components x, y) whilst inflexions and vertices appear as the zeros of a *single* smooth function in one variable (the curvature κ in the case of inflexions, and its derivative κ' in the case of vertices). We expect a general smooth function in one variable to have a discrete set of zeros, hence that inflexions or vertices will arise from a discrete set of parameters. On the other hand we do not expect two general smooth functions in one variable to have a common zero, so do not expect to have irregular parameters on a general curve z. Thus irregular parameters should be viewed as *unstable* features of a general curve z (expected to disappear under small changes in z) whereas inflexions and vertices should be viewed as *stable* features (expected to persist under small changes).

The picture alters fundamentally when we move from the study of a single curve z to the two parameter *family* of trajectories resulting from a motion μ, one for each tracing point w. Let us fix some parameter t. Consider first the condition on w for t to be irregular for the associated trajectory, represented by the common zeros of two smooth functions on the w-plane. We expect the zeros of a smooth function on a plane to be a 'curve' in some sense, so expect the common zeros of two functions to be the intersections of two 'curves', hence an isolated set of points. Indeed that is exactly what we found. (Lemma 14.1.) Provided t is generic we get a *unique* tracing point w with t irregular for the associated trajectory. In Section 15.2 we pursue this question further, and find that such parameters are always cusps: normally they are ordinary cusps, failing to be ordinary precisely when $2\kappa_p = \kappa_q$, where κ_p, κ_q are the curvatures of the fixed and moving centrodes p, q. By contrast, the condition on a general tracing point w for t to be an inflexion (or a vertex) is represented by the zeros of a single function on the w-plane, so should be a 'curve'. In Section 15.3 we study the case of inflexions. It turns out that this 'curve' is a circle, the 'inflexion circle' long familiar to mechanical engineers. Moreover, on this circle there is in principle a unique tracing point (the Ball point) for which the trajectory has an undulation at t. Finally, in Section 15.4 we study the case of vertices, where the 'curve' turns out to be a cubic curve, known to the engineering community as the 'cubic of stationary curvature'. Moreover, on this cubic there are in principle four points (the Burmester points) for which the trajectory has a higher vertex at the instant t.

15.1 Equivalence of Motions

The object of this section is to introduce natural equivalence relations on planar motions allowing us to restrict our attention to motions with very specific properties. That will enable us to simplify the theoretical computations of this chapter. Two planar motions μ_1, μ_2 with domains I_1, I_2 are *equivalent*, when there exist a change of parameter $s : I_2 \to I_1$, and fixed congruences A, B, such that for all t in I_2

$$\mu_2(t) = A\mu_1(s(t))B. \tag{15.1}$$

The reader will readily verify that 'equivalence' is indeed an equivalence relation of the set of planar motions. It is worth isolating two special cases. The motions μ_1, μ_2 are *parametrically equivalent* when there exists a change of parameter s for which (15.1) holds with $A = B = 1$, where 1 is the identity in the group of congruences: and μ_1, μ_2 are *congruent* when $I_1 = I_2$ and there exist fixed congruences A, B for which (15.1) holds with $s(t) = t$. Concerning the former concept we have the following consequence of the Uniqueness Theorem.

Lemma 15.1 *Let μ be a generic motion, and let t_0 be a fixed parameter. Then there exist a unit complex number u, and a change of parameter s with $s(t_0) = 0$, for which the rotational part $\rho(t) = ue^{is(t)}$. Thus μ is*

15.1 Equivalence of Motions

parametrically equivalent to a motion whose rotational part is given by a formula ue^{it}, defined on an interval containing $t = 0$.

Proof We need only observe that the rotational part $\rho(t)$ can be viewed as a regular curve whose trace is contained in the unit circle. Then by Lemma 6.9 there exist a unit complex number u, and a change of parameter s with $s(t_0) = 0$, for which $\rho(t) = ue^{is(t)}$. The rest is clear. □

We say that a motion μ *passes through the identity* when there exists a parameter t_0 for which $\mu(t_0) = 1$. Note first that *any motion μ_1 is congruent to a motion μ_2 passing through the identity*: in (15.1) choose $s(t) = t$, $A = 1$, choose a parameter t_0, and set $B = \mu_1(t_0)^{-1}$ to obtain a motion μ_2 for which $\mu_2(t_0) = 1$. In view of this comment we may (and henceforth will) restrict our attention to motions μ which do pass through the identity.

Example 15.1 Let p, q be regular curves with the same domain having equal speeds. We claim that the resulting roulette μ passes through the identity if and only if there exists a parameter at which p, q are equitangent. By (13.1) the rotational and translational parts of μ are given by

$$\rho = \frac{p'}{q'}, \quad \tau = \frac{pq' - p'q}{q'}.$$

The condition for μ to pass through the identity is that there exists a parameter t_0 for which $\mu(t_0) = 1$, or equivalently $\rho(t_0) = 1$, $\tau(t_0) = 0$. In view of the displayed relations that is the same as saying

$$p'(t_0) = q'(t_0), \quad p(t_0)q'(t_0) - p'(t_0)q(t_0) = 0,$$

equivalent to $p(t_0) = q(t_0)$, $p'(t_0) = q'(t_0)$, the condition for p, q to be equitangent at t_0.

Note that when two congruent motions μ_1, μ_2 pass through the identity at the same parameter t_0, the congruences A, B in (15.1) are related by $AB = 1$, so are mutually inverse: in that case we say that μ_1, μ_2 are *conjugate* motions. The concept of passing through the identity is invariant under parametric equivalence, in the sense that if a motion passes through the identity then any motion parametrically equivalent to it has the same property. Moreover Lemma 15.1 tells us that any generic motion passing through the identity is parametrically equivalent to one whose rotational part ρ is given by $\rho(t) = e^{it}$, and whose translational part τ satisfies $\tau(0) = 0$.

Lemma 15.2 *Let μ be a generic motion having rotational part $\rho(t) = e^{it}$ and translational part τ with $\tau(0) = 0$. Then μ is conjugate to a motion μ_1 whose rotational part is ρ, and whose translational part τ_1 is such that $\tau_1(0) = 0$, $\tau_1'(0) = 0$, $\tau_1''(0) = -2ir$ for some real constant $r \geq 0$.*

Proof We will find a congruence B for which $\mu_1(t) = B^{-1}\mu(t)B$ has the desired properties. Automatically, $\mu_1(0) = 1$, so $\rho_1(0) = 1$, $\tau_1(0) = 0$. Let $B(z) = uz + b$, where u, b are complex numbers with u unit. By computation

$$\rho_1 = \rho, \quad \tau_1 = u^{-1}(\rho b + \tau - b).$$

Differentiating the latter formula twice yields the relations

$$\tau_1' = u^{-1}(\rho' b + \tau'), \quad \tau_1'' = u^{-1}(\rho'' b + \tau'').$$

Setting $t = 0$ in the first relation we see that $\tau_1'(0) = 0$ if and only if $\rho'(0)b + \tau'(0) = 0$, i.e. if and only if $b = q(0)$ where q is the moving centrode of μ. That determines b, uniquely. Likewise, setting $t = 0$ in the second formula we see that an appropriate choice of u will force $\tau_1''(0)$ to have the desired form. $\square$

A generic motion μ_1 satisfying the conditions of Lemma 15.2 is said to be in *normal form*. Thus we have established that *any generic motion is equivalent to one in normal form*. The next result spells out the consequences for the centrodes.

Lemma 15.3 *The centrodes p, q of a generic motion μ in normal form have the properties that $p(0) = q(0) = 0$, and that $p'(0) = q'(0) = 2r$ for some real constant $r \geq 0$. Further, they fail to be regular at $t = 0$ if and only if $\tau''(0) = 0$.*

Proof The centrodes p, q are defined by the relations $p = \rho q + \tau$, $\rho' q + \tau' = 0$. Setting $t = 0$ gives $p(0) = 0$, $q(0) = 0$. Differentiating, and then setting $t = 0$, we get $q'(0) = i\tau''(0)$, $p'(0) = q'(0)$. The result follows. $\square$

In Section 15.3 we will see that the constant r appearing in the normal form has a satisfying geometric interpretation. For the moment we content ourselves with the observation that r can be expressed neatly in terms of the curvatures of the centrodes.

Lemma 15.4 *Let μ be a generic motion in normal form with $t = 0$ regular for the centrodes p, q. Write r_p, r_q for the radii of curvature of p, q at that parameter. Then the constant r is determined by the Euler–Savary equation*

$$\frac{1}{r_p} - \frac{1}{r_q} = \frac{1}{2r}. \tag{15.2}$$

Proof Differentiate the identity $p = \rho q + \tau$ twice, and then set $t = 0$ to get $p''(0) - q''(0) = 2ir$, hence $\Im p''(0) - \Im q''(0) = 2r$. By (5.5) the curvatures κ_p, κ_q of p, q at $t = 0$ are determined by $4r^2\kappa_p = \Im p''(0)$, $4r^2\kappa_q = \Im q''(0)$. The result follows. □

The main point to make before proceeding is that the concepts (of cusps, inflexions and vertices on trajectories) studied in this chapter are invariant under the notion of equivalence, so can be studied by working with an equivalent motion. The details are spelled out in the following exercises.

Exercises

15.1.1 Show that the relations of 'equivalence', 'parametric equivalence' and 'congruence' on motions are equivalence relations.

15.1.2 Let μ_1, μ_2 be parametrically equivalent motions under a change of parameter s. Show that the associated fixed centrodes p_1, p_2, and the associated moving centrodes q_1, q_2, are parametrically equivalent under the same change of parameter s.

15.1.3 Let μ_1, μ_2 be parametrically equivalent motions under a change of parameter s. Show that for any tracing point w the trajectory of w under μ_1 is parametrically equivalent to the trajectory under μ_2 via the same change of parameter s.

15.2 Cusps on Trajectories

It is time to say something about the *nature* of irregular points of trajectories at generic parameters. Given a motion μ, and a generic parameter t, we know that there is a unique tracing point w with the property that t is irregular for the trajectory under μ. The next lemma tells us considerably more: it should be read in the light of the general intuition explained in Section 14.1.

Lemma 15.5 *Let μ be a generic motion, and let t be a regular parameter for the centrodes p, q. Then t is a cusp for the trajectory of the moving instantaneous centre $w = q(t)$, failing to be an ordinary cusp if and only if $2\kappa_p = \kappa_q$ where κ_p, κ_q are the curvatures of p, q at t.*

Proof By Lemma 15.3 we can assume μ is in normal form, so $t = 0$, $p(0) = q(0) = 0$, and $p'(0) = q'(0) = 2r > 0$. Thus $w = q(0) = 0$, and $\mu(t)(w) = \tau(t)$. The rotational part $\rho(t) = e^{it}$, so $\rho(0) = 1$, $\rho'(0) = i$, $\rho''(0) = -1$. The moving centrode q satisfies the identity $\rho'q + \tau' = 0$: differentiating twice, and setting $t = 0$, we obtain

$$\tau'(0) = 0, \quad -\tau''(0) = 2ir, \quad -\tau'''(0) = -4r + iq''(0). \tag{15.3}$$

In particular $\tau''(0) \neq 0$, so $t = 0$ is a cuspidal parameter. It fails to be ordinary if and only if $\tau''(0)$, $\tau'''(0)$ are linearly dependent, i.e. if and only if $\Im q''(0) = -4r$. However, the proof of Lemma 15.4 shows that $\Im q''(0) = 4r^2\kappa_q$, so the condition is $\kappa_q = -1/r$. By the Euler–Savary equation (15.2) that holds if and only if $2\kappa_p = \kappa_q$. $\square$

Here are two applications of this result, illustrating situations where we can show that higher cusps cannot occur on the basis of the underlying geometry.

Example 15.2 The first application is to involutes of a regular curve p, thought of as the trajectories of the roulette obtained by rolling a tangent line q along the curve. (Example 13.5.) We claim that *a generic parameter t must be an ordinary cusp* of the trajectory for $q(t)$. Suppose t is a higher cusp. Lemma 15.5 then tells us that $2\kappa_p = \kappa_q$, and since $\kappa_q = 0$ it would follow that $\kappa_p = 0$, i.e. that t is inflexional for p at t: but this contradicts the result of Example 14.3 telling us that t fails to be generic if and only if t is inflexional for p.

Example 15.3 The second application is to orthotomics of a regular curve p, thought of as the trajectories of the roulette obtained by rolling a reflexion q (of p in some tangent line) along p. (Example 13.6.) Again, we claim that *a generic parameter t must be an ordinary cusp* of the trajectory for $q(t)$. For suppose t is a higher cusp. Lemma 15.5 then gives $2\kappa_p = \kappa_q$, and since $\kappa_q = -\kappa_p$ we deduce that $\kappa_p = \kappa_q = 0$, i.e. that t is inflexional for both p, q: but this contradicts the result of Example 14.4 telling us that then t is non-generic.

15.3 Inflexions on Trajectories

Finally, it is worth remarking that the appearance of non-generic parameters ($\kappa_p = \kappa_q$), and that of higher cusps ($2\kappa_p = \kappa_q$), are *stable* phenomena, meaning that they persist under small changes in the centrodes p, q. Indeed, small changes in p, q will lead only to small changes in $2\kappa_p - \kappa_q$ and $\kappa_p - \kappa_q$: it remains to observe that in general a zero of a smooth function persists under small deformations.

15.3 Inflexions on Trajectories

Following the introduction to this chapter we now study the question of describing inflexions on the trajectories of a motion μ. Thus we fix some parameter t, and consider the set of tracing points w with the property that t is inflexional for the trajectory of w under μ: interestingly, it turns out to be a circle.

Lemma 15.6 *Let μ be a generic motion, let t be a regular parameter for the centrodes p, q, and let $J(t)$ be the set of all tracing points w for which t is inflexional for the trajectory of w. Then $J(t)$ is a circle through $q(t)$ with the point $q(t)$ deleted. The circle is tangent to q at $q(t)$, with radius r determined by the Euler–Savary equation (15.2).*

Proof By Lemma 15.3 we can assume μ is in normal form. Thus $t = 0$, the rotational part ρ satisfies $\rho(0) = 1$, $\rho'(0) = i$, $\rho''(0) = -1$, and the translational part τ satisfies $\tau(0) = 0$, $\tau'(0) = 0$, and $\tau''(0) = -2ir$ for some positive real constant r. We seek those tracing points w for which the first two derivatives $\rho'(0)w + \tau'(0)$, $\rho''(0)w + \tau''(0)$ of the trajectory are linearly dependent, i.e. for which iw, $-w - 2ir$ are linearly dependent. Setting $w = u + iv$, this condition reduces to $u^2 + v^2 + 2rv = 0$, defining the circle of radius r centre $(0, -r)$. □

The circle containing $J(t)$ is the *moving inflexion circle*, mapped under $\mu(t)$ to another circle, the *fixed inflexion circle*, of the same radius r passing through $p(t)$ and tangent there to p: we write $I(t)$ for the set obtained by deleting p from this circle. Either circle is determined by a single inflexion: indeed a circle tangent to a fixed line at a fixed point on that line is completely determined by any one of its points. Lemma 15.6 interprets the constant r which appeared in the Euler–Savary equation as the common radius of the inflexion circles.

Example 15.4 Consider again the double slider of Example 14.5. Recall that the centrodes are Cardan circles, and that the trajectory of a point

w is an ellipse, collapsing to a diameter of the rolling circle when w lies on that circle. For this example we can determine the inflexion circle without further computations. Since ellipses have no inflexions, the only trajectories which have inflexions are the diameters (for which *every* point is an inflexion, bar the extremities) which are the trajectories of points on the rolling circle. It follows that *the moving inflexion circle at any given instant t is the rolling circle*. We can verify our conclusion by direct calculation, following the proof of Lemma 15.6. Recall that the rotational and translational parts of the motion are $\rho(t) = e^{-it}$, $\tau(t) = i\sin t$. Then $\rho'(0) = -i$, $\tau'(0) = i$, $\rho''(0) = -1$, $\tau''(0) = 0$ and the moving inflexion circle is $u^2 + v^2 - u = 0$, agreeing with the equation for the rolling circle in Example 14.5.

Example 15.5 Consider the inflexions of the limacons obtained by rolling one circle q on another fixed circle p having the same radius r. For calculations take

$$p(t) = ri(1 - e^{it}), \quad q(t) = -ri(1 - e^{-it}).$$

Thus $p(t)$ parametrizes the circle of radius r centred at $(0, r)$, and $q(t)$ the circle of radius r centred at $(0, -r)$, obtained from p by reflexion in the common tangent line $y = 0$. Note that p, q have equal speeds, and are equitangent at $t = 0$. With these choices the resulting roulette μ is easily checked to have rotational and translational parts

$$\rho(t) = e^{2it}, \quad \tau(t) = ri(e^{it} - 1)^2.$$

We will find the inflexion circle $J(0)$. We have $\rho'(0) = 2i$, $\tau'(0) = 0$, $\rho''(0) = -4$, $\tau''(0) = -2ir$, so $J(0)$ has equation $2u^2 + 2v^2 + rv = 0$, a circle of radius $r/4$ centre $(0, -r/4)$. That agrees with the conclusion of Example 5.9, that inflexional limacons are the trajectories of points inside the rolling circle and nearer to the circumference than the centre.

Example 15.6 The previous example is a special case of the general situation for orthotomics, thought of as the trajectories of roulettes generated by regular curves p, q where q is the reflexion (in some tangent line) of p. Such curves will have curvatures κ_p, κ_q related by $\kappa_p = -\kappa_q$. According to Example 14.4 a generic parameter t is one which is not inflexional for p. And Lemma 15.6 tells us that the inflexion circle at such a parameter will have radius $1/4|\kappa_p|$: in other words the radius of the inflexion circle is one quarter the radius of curvature r_p of p.

15.3 Inflexions on Trajectories

Example 15.7 Another example is provided by involutes, thought of as trajectories of roulettes generated by regular curves p, q where q is a tangent line to p. According to Example 14.3 a generic parameter t is one which is not inflexional for p. In this example $\kappa_q = 0$, and Lemma 15.6 tells us that for a generic parameter t the radius of the inflexion circle is one half the radius of curvature r_p of p.

Example 15.8 A special case arises when the moving inflexion circle coincides with the circle of curvature of the moving centrode q. It is interesting to see how this relates to our discussion of cusps. The condition for the circles to coincide is that $r = -1/\kappa_q$. As we saw in the proof of Lemma 15.5 that is precisely the condition for the trajectory of the instantaneous moving centre to fail to have an ordinary cusp at $t = 0$, equivalently that $2\kappa_p = \kappa_q$.

It is natural to push the inflexion question a little further by asking about *undulations* of trajectories. The situation so far is that for a roulette μ we know that at a generic parameter t all the tracing points w for which t is inflexional for the associated trajectory lie on the moving inflexion circle $J(t)$. Intuitively, one would expect a discrete set of tracing points w on $J(t)$ for which t is undulational for the trajectory. We can be more precise.

Lemma 15.7 *Let μ be a generic planar motion, and let t be a regular parameter for the centrodes p, q. In general there is just one tracing point w on $J(t)$ with t undulational for the trajectory. An exceptional case arises when $2\kappa_p(t) = \kappa_q(t)$: then, either no points on $J(t)$ give rise to undulations, or every point does.*

Proof By Lemma 15.3 we can assume μ is in normal form. Thus $t = 0$, the rotational part ρ is given by $\rho(t) = e^{it}$, and by (15.3) the translational part τ satisfies

$$\tau'(0) = 0, \quad \tau''(0) = -2ir, \quad \tau'''(0) = 4r - iq''(0).$$

The points on $J(0)$ correspond to those coupler points w for which the first two derivatives $\rho'(0)w + \tau'(0)$, $\rho''(0)w + \tau''(0)$ of the trajectory are linearly dependent. By Section 7.3 the condition for t to be undulational for the trajectory is that in addition the first and third derivatives $\rho'(0)w + \tau'(0)$, $\rho'''(0)w + \tau'''(0)$ should be linearly dependent. Setting $w = u + iv$, $\tau'''(0) = a + ib$ this condition reduces to $au + bv = 0$. The general case is when $a \neq 0$:

the condition then defines a line through the origin in the w-plane, intersecting the moving inflexion circle in just one further point, as claimed. It remains to consider the exceptional case $a = 0$. When $b \neq 0$ our line is the tangent $v = 0$ to the circle at the origin, so there are no further intersections with the circle, and we obtain no undulations: and when $b = 0$ the condition $au + bv = 0$ is satisfied by every tracing point w, and hence every point on $J(0)$ gives rise to an undulation. It remains to interpret the conditions. The condition $a = 0$ for the exceptional case is equivalent to $\Im q''(0) = -4r$. As in the proof of Lemma 15.5 that holds if and only if $2\kappa_p = \kappa_q$. $\square$

Here are two examples, illustrating Lemma 15.7 for two very familiar motions.

Example 15.9 In Example 15.5 we considered the family of limacons obtained by rolling the circle q of radius r and centre $(0, -r)$ on the fixed circle p of radius r and centre $(0, r)$, and discovered that the moving inflexion circle at $t = 0$ is the circle $2u^2 + 2v^2 + rv = 0$ of radius $r/4$ centred at the point $(0, -r/4)$ where $w = (u, v)$ is the tracing point. The proof of Lemma 15.7 shows that $t = 0$ is undulational for the trajectory of w if and only if $au + bv = 0$ where $\tau'''(0) = a + ib$ with a, b real. By computation, $\tau'''(0) = 6r$ so $a = 6r$, $b = 0$ and the condition reduces to $u = 0$, a line intersecting the inflexion circle in a unique point $(0, -r/2)$. That agrees with the calculation of Example 3.14 where we saw that the undulational limacon was the trajectory of a point midway between the centre and the circumference of the rolling circle.

Example 15.10 Recall that for the double slider of Example 14.5 with $\rho(t) = e^{-it}$, $\tau(t) = i \sin t$, the fixed centrode p has trace the circle of radius 1 centred at $(0, 0)$, the moving centrode q has trace the circle of radius $1/2$ centred at the point $(1/2, 0)$, and the trajectories of the resulting roulette are generally ellipses, possibly collapsing to a line segment. Example 15.4 verified that the moving inflexion circle coincides with the moving centrode. Since the only inflexions occur on trajectories which are line segments, and these are necessarily undulations, we expect *every* point on the inflexion circle at $t = 0$ to give rise to an undulation. That is easily verified: simply note that $\rho''' = -\rho'$, $\tau''' = -\tau'$ so that for any point w the first and third derivatives of the trajectory $\rho w + \tau$ are linearly dependent. In this example the 'line' in the proof of Lemma 15.7 degenerates to the whole plane.

15.4 Vertices on Trajectories

The import of Lemma 15.7 is that for a *generic* planar motion we expect to find a unique undulational tracing point $w(t)$ for each generic parameter t, defining a curve in the moving plane, known to kinematicians as the *Ball curve*. For instance in Example 15.5 it is clear that the trace of the Ball curve is the circle of radius $3r/2$ concentric with the fixed circle. We will not pursue the geometry of the Ball curve any further, save to point out that in principle there should be a discrete set of parameters t for which $w(t)$ corresponds to *higher* undulations of trajectories, i.e. points where the tangent line has at least 5-point contact. The Watt four bar gives rise to a motion having exactly one such point, and is the progenitor of a number of interesting linkages yielding approximate straight line motion.

15.4 Vertices on Trajectories

Lemma 15.8 *Let μ be a generic motion, and let t be regular for the centrodes p, q. Write $K(t)$ for the set of all tracing points w with the property that t is a vertex for the trajectory of w. Then $K(t)$ is a cubic curve through the moving instantaneous centre $q(t)$, with the point $q(t)$ deleted.*

Proof By Lemma 15.3 we can assume μ is in normal form. Thus $t = 0$, the rotational part ρ is given by $\rho(t) = e^{it}$, and by (15.3) the translational part τ satisfies

$$\tau'(0) = 0, \quad \tau''(0) = -2ir, \quad \tau'''(0) = 4r - iq''(0).$$

Recall that vertices are stationary parameters t of the curvature function κ. The relation (5.5) gave a formula for the curvature (at a regular parameter t) of a plane curve z with components x, y: differentiating that formula we see that vertices appear if and only if

$$(x'^2 + y'^2)(x'y''' - x'''y') = 3(x'x'' + y'y'')(x'y'' - x''y'). \tag{15.4}$$

We are concerned with the case when z is the trajectory of w, so given by $z(t) = \rho(t)w + \tau(t)$. For calculations set $w = u + iv$, $\tau'''(0) = a + ib$ where u, v, a, b are all real. Differentiating z successively, and setting $t = 0$ in the derivatives, we get

$$x(0) = u, \quad x'(0) = -v, \quad x''(0) = -u, \quad x'''(0) = v + a$$
$$y(0) = v, \quad y'(0) = u, \quad y''(0) = -v - 2r, \quad y'''(0) = -u + b.$$

Setting $t = 0$ in (15.4) we see that $K(0)$ is defined by the following relation, which is in principle a cubic curve in the (u, v)-plane passing through the origin.

$$(u^2 + v^2)\{(a - 6r)u + bv\} = 12r^2 uv. \tag{15.5}$$

□

The cubic defined by (15.5) is known as the *cubic of stationary curvature*. The reader with a working knowledge of algebraic curves will enjoy analysing its geometry. In general it is irreducible with a crunode at the origin, whose tangents are the tangent and normal lines $v = 0$, $u = 0$ to the moving centrode: moreover it is a 'circular' cubic, in the sense that the corresponding cubic in the complex projective plane passes through the circular points at infinity $I = (1 : i : 0)$, $J = (1 : -i : 0)$. Naturally, one would like to go further and ask for those tracing points w which give rise to *higher* inflexions on the trajectory. These turn out to be the intersections of the cubic of stationary curvature with a second circular cubic having a node at the origin. In principle these cubics intersect in nine points: two are at I, J and three at the common node, leaving four points in the Euclidean plane known as the *Burmester points*. And as the parameter t varies so these four Burmester points will themselves describe interesting curves, whose geometry will reflect the intimate detail of the motion.

That brings the material of this book to a natural conclusion, with differential geometry merging imperceptibly into algebraic geometry. There is much intriguing material to be developed here. However, that is another story, reserved for those with a serious interest in kinematics. Our objective has been achieved.

Index

additive, 52
Agnesi's versiera, 39
Alabone, xvii
algebraic, 40
algebraic curve, 38
amplitude, 55
angle, 4
angular velocity, 55
antiorthotomic, 163
arc, 16
arc length, 50
arch of cycloid, 48
astroid, 44, 48, 51, 52, 112, 120, 142, 149, 161, 179
asymptote, 35
auxiliary circle, 34
axis
   of ellipse, 34
   of hyperbola, 35
   of parabola, 34

Ball curve, 209
bar, 182
bar length, 182
basis
   orthogonal, 5
   orthonormal, 5
Bernoulli, 16, 48
   lemniscate, 158, 179
Bézier, 26
Bézier curve, 26
biflexional limaçon, 98
bilinear, 2
brachistochrone, 48
branch
   of cross curve, 28, 72
   of hyperbola, 36, 54
bullet nose, 38
Burmester point, 210

Cardan circle, 46, 197
cardioid, 44, 46, 48, 52, 71, 117, 132, 144, 161, 169, 197
catenary, 16, 52, 61, 117, 131, 177
Cauchy Inequality, 3
caustic, 166
   of circle, 171, 176
   of ellipse, 172
   of equiangular spiral, 172
   of parabola, 176
Cayley sextic, 21, 52, 53, 117, 119, 159
central dilation, 79
central reflexion, 81
centre, 79
   of circle, 14
   of ellipse, 34
   of hyperbola, 35
   of rotation, 80
   of trochoid, 42
change of parameter, 54
Chasles, 190
   Theorem, 197
circle, 14, 28, 110
   arc of a, 31
   auxiliary, 34
   Cardan, 46
   concentric, 15
   fixed, 42
   inflexion, 205
   moving, 42
   of curvature, 109
   unit, 14
cissoid of Diocles, 39, 73, 104, 118, 155
clover leaf, 17
coefficients, 38
coincident lines, 11
complex conjugate, 6
component, 13
   orthogonal, 2
   parallel, 2

concentric circles, 15
condition
 tangency, 139
 variability, 138
congruence, 80
 of motions, 200
congruent
 rotationally, 82
 translationally, 82
conic, 32, 39
constant curve, 15
contact
 even, 92
 function, 92, 105
 infinite, 92
 k point, 92, 106
 odd, 92
control point, 26
convex, 133
coordinates, 1
 standard, 2
cosine rule, 5
coupler plane, 182
crank, 182
cross curve, 28, 72
crossed parallelogram, 195
cubic, 39
 of stationary curvature, 210
 Tschirnhausen, 20
curtate cycloid, 48
curvature, 64
 centre of, 109
 circle of, 109
 limiting, 73, 102, 104
 of catenary, 117
 of circle, 66
 of cycloid, 68
 of eight-curve, 72
 of ellipse, 66
 of equiangular spiral, 73
 of graph, 66
 of hyperbola, 130
 of limacon, 68
 of line, 66
 of orthotomic, 159
 of parabola, 68
 radius of, 109
curve
 algebraic, 38, 40
 Ball, 209
 Bézier, 26
 bullet nose, 38
 clover leaf, 17
 constant, 15
 convex, 133
 cross, 28, 72
 eight-, 20, 92, 99, 112, 129

equitangent, 187
equivalent, 84
 orthotomic, 153, 174
 parallel, 119, 131
 parametrized, 13
 Peano, 13
 pedal, 159
 periodic, 17, 133
 regular, 24
 rose, 17, 55, 85, 106
 segment, 16
 unit speed, 69
cusp, 100
 higher, 104
 of cardioid, 132
 of cissoid, 104
 of cycloid, 104
 of epicycloid, 104
 of evolute, 131
 of hypocycloid, 104
 of involute, 204
 of limacon, 101
 of Lissajous figure, 104
 of orthotomic, 204
 of parallel, 131
 of piriform, 103
 of semicubical parabola, 99–101
 of tractrix, 104
 of trajectory, 203
 ordinary, 102
cuspidal cycloid, 48, 68
cuspidal parameter, 100
cycloid, 47, 49, 60, 99, 104, 130, 188
 curtate, 48
 cuspidal, 48, 68
 prolate, 48
cycloidal pendulum, 61

degree, 38
degree of freedom, 181
deltoid, 44, 161, 179
Descartes' folium, 41, 73
de Casteljau, 26
Diocles, 40
direct isometry, 80
directrix, 32, 163
distance, 4, 12
dot product, 2
double cardioid, 44
double slider, 185, 205, 208

eccentricity, 33
eight-curve, 20, 92, 112, 129
ellipse, 33, 34, 37, 38, 46, 68, 111, 127, 142, 149, 159, 172, 195, 197
 standard, 34
 standard parametrization, 34

envelope, 138
Envelope Theorem, 140
epicycloid, 44, 114, 144
epitrochoid, 43
equiangular spiral, 58, 73, 82, 118, 172
equidistant, 10
equitangent curves, 187
equivalent curves, 84
Euclidean group, 77
Euclidean structure, 2
Euler–Savary equation, 203
evolute, 109, 148, 149
 of cardioid, 132
 of catenary, 118
 of circle, 110
 of cycloid, 116
 of ellipse, 111, 118
 of epicycloid, 114, 118
 of equiangular spiral, 118
 of graph, 119
 of hypocycloid, 114, 118
 of parabola, 111, 124
 of semicubical parabola, 112
 of tractrix, 117
Existence Theorem, 75

Factor Theorem, 90, 91, 104
family
 one parameter, 137
fixed
 centrode, 194
 circle, 42
 inflexion circle, 205
 instantaneous centre, 192
 plane, 42, 186
 point, 79
focus, 32
Formula
 Serret–Frenet, 63
four bar linkage, 182
Four Vertex Theorem, 135
frame, 62
 moving, 62

Galileo, 16
Geminus, 40
generic motion, 194
generic parameter, 191
gradient, 99
 limiting, 99
Grashof type, 182
group
 Euclidean, 77

half line, 17
half term, 81
half turn, 82

higher vertex, 126
Huyghens, 61
hyperbola, 33, 35, 36, 38, 57, 118, 130, 158, 195
 rectangular, 35
 standard, 35
hypocycloid, 44, 114, 144
hypotrochoid, 43, 55

incident ray, 166
indirect isometry, 80
infinite contact, 92
inflexion, 69, 95, 96
 circle, 205
 of cissoid of Diocles, 73
 of cross curve, 72
 of cycloid, 99
 of Descartes' folium, 73
 of eight-curve, 72, 99
 of epicycloid, 71
 of equiangular spiral, 73
 of graph, 69, 96
 of hypocycloid, 71
 of involute, 207
 of Kampyle, 73
 of limacon, 71, 98, 171, 206, 208
 of orthotomic, 206
 of piriform, 70
 of right strophoid, 73
 of Serpentine, 97
 of trajectory, 205
 of versiera, 70
 ordinary, 95, 96
inflexional parameter, 69
initial time, 184
Intermediate Value Theorem, 134
inverse motion, 198
involute, 59, 114, 207
 as roulette, 188
 backward, 61
 forward, 61
 of catenary, 61
 of circle, 59
 of cycloid, 60
irregular parameter, 24
isochronous, 61
isometry, 77
 linear part of, 81
 direct, 80
 indirect, 80

Jacobian matrix, 140

Kampyle of Eudoxus, 73

length, 3

limacon, 46, 68, 71, 98, 101, 129, 159, 206, 208
  biflexional, 72, 98
  nodal, 47
  undulational, 72, 98
  uniflexional, 98
line(s), 7, 39
  coincident, 11
  direction, 9
  half, 17
  intersecting, 11
  normal, 27
  orthogonal, 10
  parallel, 11
  parametrized, 8
  segment, 17, 30, 31, 85
  tangent, 27
linearly independent, 1
linkage
  collapsible, 183
  four bar, 182
Lissajous figure, 55, 104
Lockwood, xvii

Maclaurin's trisectrix, 23
members, 137
mid-point, 10
mirror, 166
modulus, 6
motion
  congruent, 200
  conjugate, 201
  equivalent, 200
  generic, 194
  in normal form, 202
  inverse, 198
  parametrically equivalent, 200
  planar, 184
  smooth, 184
moving
  centrode, 194
  circle, 42
  frame, 62
  inflexion circle, 205
  instantaneous centre, 192
  plane, 42, 186
Mukhopadhaya, 134

nephroid, 44, 117, 119, 142, 167, 175, 176
Newton formula, 69
nodal limacon, 47
normal
  form of a motion, 202
  line, 27
  unit, 62
  vector, 23
Norwich spiral, 60

ordinary
  vertex, 126
orthogonal
  bisector, 10, 151
  line, 10
  projection, 12
  vector, 2
orthotomic, 153, 174, 206
  as roulette, 188
  of astroid, 161
  of cardioid, 159, 161
  of circle, 154, 159, 175, 179
  of deltoid, 161
  of ellipse, 157, 159
  of epicycloid, 160
  of hyperbola, 158, 179
  of hypocycloid, 160
  of line, 174
  of parabola, 155, 156, 159, 175
  of starfish, 161
oval, 133

parabola, 27, 33, 37, 53, 68, 107, 111, 124, 126, 127, 148, 156, 159, 163, 175, 176
  semicubical, 24, 52, 61
  standard, 34
  standard parametrization, 34
parallel
  curve, 119
  lines, 11
  of circle, 123
  of line, 123
  of parabola, 120, 123
parallelogram
  crossed, 195
parallelogram law, 5
parameter, 13
  change of, 54
  generic, 191
  inflexional, 69
  irregular, 24
  regular, 24
  starting, 59
parametric equivalence, 119
  of curves, 54
  of motions, 200
parametrized
  curve, 13
  line, 8
  set, 15
part
  imaginary, 6
  real, 6
pedal curve, 159
pencil of lines, 17, 39, 41, 89, 167, 173, 177
period, 17
periodic curve, 17

Index

phase constant, 55
piriform, 24, 70, 103, 130
planar
  mapping, 77
  motion, 184
plane
  fixed, 42, 186
  moving, 42, 186
point, 1
  Burmester, 210
  control, 26
  fixed, 79
  tracing, 43, 182, 187
polar form, 7
Polarization Identity, 6
positive definite, 2
pre-envelope, 138
preserves distance, 77
product, 6
prolate cycloid, 48
Pythagoras Theorem, 5

quartic, 39

radius
  of circle, 14
  of curvature, 109
ratio, 99
reflected ray, 166
reflective property
  for a hyperbola, 172
  for a parabola, 176
  for an ellipse, 172
reflexion, 152
  central, 81
  of a curve, 153
regular
  curve, 24
  parameter, 24
reparametrization, 54
rhodonea, 17
right strophoid, 15, 73, 156
Rolle's Theorem, 128
rose curve, 17, 22, 24, 55, 85, 106
rotation, 80
rotational part, 184
roulette, 42, 187

scalar, 1
scalar product, 2
scaling factor, 79
self crossing, 21
semiaxis
  major, 34
  minor, 34
semicubical parabola, 24, 52, 61, 99–101, 103, 112

Serpentine, 97, 131
Serret–Frenet Formulas, 63
similarity, 79
simple harmonic motion, 55
singular set, 140
smooth motion, 184
source, 151, 153, 166
special Euclidean group, 80
speed, 50
spiral
  equiangular, 73, 82, 172
  Norwich, 60
  pole of, 82
spirograph, xvii, 42
standard
  basis, 1, 5
  coordinates, 2
  ellipse, 34
  equiangular spiral, 83
  hyperbola, 35
  parabola, 34
starfish, 44, 161
starting parameter, 59
strophoid
  right, 156
strophoid, right, 15
symmetry, 2

tangency condition, 139
tangent, 106
  line, 27
    cuspidal, 100
    limiting, 99
  unit, 62
  vector, 23
    cuspidal, 100
trace, 15
tracing point, 43, 182, 187
tractrix, 28, 61, 104, 117
trajectory, 182, 184
translation, 77
translational part, 184
Triangle Inequality, 3, 4
trisectrix of Maclaurin, 23
trochoid, 42, 49, 187
Tschirnhausen cubic, 20, 27

undulation, 95, 98, 181, 207
  of limacon, 208
  of trajectory, 207
uniflexional limacon, 98
Uniqueness Lemma, 66
Uniqueness Theorem, 87
unit
  speed, 57
  vector, 4

variability condition, 138
vector, 1
  normal, 23
  orthogonal, 2
  tangent, 23
versiera, 39, 57, 70, 131
vertex, 126
  higher, 126
  of cardioid, 132
  of catenary, 131
  of cycloid, 130
  of eight-curve, 129
  of ellipse, 34, 127
  of graph, 128, 130
  of hyperbola, 35, 130
  of limacon, 129
  of oval, 135
  of parabola, 34, 126, 127
  of piriform, 130
  of Serpentine, 131
  of trajectory, 209
  of versiera, 131
  ordinary, 126

Watt four bar, 181

zero, 89
  k-fold, 91
  of finite multiplicity, 91
  of infinite multiplicity, 91
  repeated, 91
zero set, 40